VOLUME ONE HUNDRED AND FIFTY NINE

PROGRESS IN MOLECULAR BIOLOGY AND TRANSLATIONAL SCIENCE

VOLUME ONE HUNDRED AND FIFTY NINE

PROGRESS IN MOLECULAR BIOLOGY AND TRANSLATIONAL SCIENCE

Edited by

DAVID B. TEPLOW
Department of Neurology,
David Geffen School of Medicine at UCLA,
Los Angeles, CA, United States

ELSEVIER

ACADEMIC PRESS
An imprint of Elsevier

Academic Press is an imprint of Elsevier
50 Hampshire Street, 5th Floor, Cambridge, MA 02139, United States
525 B Street, Suite 1650, San Diego, CA 92101, United States
The Boulevard, Langford Lane, Kidlington, Oxford OX5 1GB, United Kingdom
125 London Wall, London EC2Y 5AS, United Kingdom

First edition 2018

ISBN: 978-0-12-816235-4
ISSN: 1877-1173

Publisher: Zoe Kruze
Acquisition Editor: Ashlie M. Jackman
Editorial Project Manager: Joanna Collett
Production Project Manager: James Selvam
Cover Designer: Matthew Limbert

Typeset by SPi Global, India

CONTENTS

Contributors *ix*
Preface *xi*

1. Targeting the Recently Deorphanized Receptor GPR83 for the Treatment of Immunological, Neuroendocrine and Neuropsychiatric Disorders **1**

Lindsay M. Lueptow, Lakshmi A. Devi, and Amanda K. Fakira

1. Introduction 2
2. Expression and Significance of GPR83 in the Brain 7
3. Role of GPR83 in Immune Function 15
4. Current Understanding of GPR83 and PEN 18
5. Conclusions 19

Acknowledgment 21
References 21

2. Arrestins in the Cardiovascular System: An Update **27**

Anastasios Lymperopoulos

1. Introduction 28
2. Cardiovascular Adrenergic Receptors and βarrestins 29
3. Cardiovascular Angiotensin II Receptors and βarrestins 37
4. Other Cardiovascular GPCRs and βarrestins 42
5. Therapeutic Implications of the Functional Divergence of Cardiovascular βarrestins 47
6. Conclusions and Future Perspectives 49

References 50

3. Global Aquatic Hazard Assessment of Ciprofloxacin: Exceedances of Antibiotic Resistance Development and Ecotoxicological Thresholds **59**

Kaitlyn R. Kelly and Bryan W. Brooks

1. Background 60
2. Materials and Methods 62
3. Results and Discussion 63
4. Conclusions 74

References 75

4. Group I Intron-Based Therapeutics Through *Trans*-Splicing Reaction **79**

Chang Ho Lee, Seung Ryul Han, and Seong-Wook Lee

1. Introduction 80
2. Group I Intron 82
3. Group I Intron as Therapeutics 86
4. Concluding Remarks 93

Acknowledgments 94

References 94

5. Major 3′–5′ Exoribonucleases in the Metabolism of Coding and Non-coding RNA **101**

Ricardo F. dos Santos, Ana P. Quendera, Sofia Boavida, André F. Seixas, Cecília M. Arraiano, and José M. Andrade

1. Introduction 102
2. Polynucleotide Phosphorylase 107
3. RNase II 121
4. RNase R 126
5. Concluding Remarks 138

Acknowledgments 140

References 140

6. Different Methods of Delivering CRISPR/Cas9 Into Cells **157**

Arun Pandian Chandrasekaran, Minjung Song, Kye-Seong Kim, and Suresh Ramakrishna

1. Introduction 158
2. Delivery Methods of CRISPR/Cas9 for Genome Editing 162
3. Opportunities and Challenges in CRISPR/Cas9 Delivery to Stem Cells 169
4. Conclusions and Future Perspectives 170

Acknowledgments 171

References 171

7. Structural Simplicity and Mechanistic Complexity in the Hammerhead Ribozyme **177**

Sara M. O'Rourke and William G. Scott

1. Background and Structural Overview 178
2. Fast Minimal Hammerhead Ribozymes 181

3. Acid-Base Catalysis and the Hammerhead Ribozyme 182
4. Is the Hammerhead Ligation Reaction the Reverse of the Cleavage Reaction? 191
5. Do Cooperative Interactions in the Hammerhead Ribozyme Facilitate General Base Catalysis in the Cleavage Reaction? 194
6. Summary and Concluding Remarks 195
References 201

Index *203*

CONTRIBUTORS

José M. Andrade
Instituto de Tecnologia Química e Biológica António Xavier, Universidade Nova de Lisboa, Oeiras, Portugal

Cecília M. Arraiano
Instituto de Tecnologia Química e Biológica António Xavier, Universidade Nova de Lisboa, Oeiras, Portugal

Sofia Boavida
Instituto de Tecnologia Química e Biológica António Xavier, Universidade Nova de Lisboa, Oeiras, Portugal

Bryan W. Brooks
Environmental Health Science Program, Department of Environmental Science; Institute of Biomedical Studies, Baylor University, Waco, TX, United States

Arun Pandian Chandrasekaran
Graduate School of Biomedical Science and Engineering, Hanyang University, Seoul, South Korea

Lakshmi A. Devi
Department of Pharmacological Sciences, Freidman Brain Institute, Icahn School of Medicine at Mount Sinai, New York, NY, United States

Ricardo F. dos Santos
Instituto de Tecnologia Química e Biológica António Xavier, Universidade Nova de Lisboa, Oeiras, Portugal

Amanda K. Fakira
Department of Pharmacological Sciences, Freidman Brain Institute, Icahn School of Medicine at Mount Sinai, New York, NY, United States

Seung Ryul Han
Rznomics Inc., Gwangju, Republic of Korea

Kaitlyn R. Kelly
Environmental Health Science Program, Department of Environmental Science, Baylor University, Waco, TX, United States

Kye-Seong Kim
Graduate School of Biomedical Science and Engineering; College of Medicine, Hanyang University, Seoul, South Korea

Chang Ho Lee
Department of Integrated Life Sciences, Dankook University, Yongin, Republic of Korea

Seong-Wook Lee
Department of Integrated Life Sciences, Dankook University, Yongin; Rznomics Inc., Gwangju, Republic of Korea

Lindsay M. Lueptow
Department of Pharmacological Sciences, Freidman Brain Institute, Icahn School of Medicine at Mount Sinai, New York, NY, United States

Anastasios Lymperopoulos
Laboratory for the Study of Neurohormonal Control of the Circulation, Department of Pharmaceutical Sciences, Nova Southeastern University College of Pharmacy, Fort Lauderdale, FL, United States

Sara M. O'Rourke
The Center for the Molecular Biology of RNA, University of California at Santa Cruz, Santa Cruz, CA, United States

Ana P. Quendera
Instituto de Tecnologia Química e Biológica António Xavier, Universidade Nova de Lisboa, Oeiras, Portugal

Suresh Ramakrishna
Graduate School of Biomedical Science and Engineering; College of Medicine, Hanyang University, Seoul, South Korea

William G. Scott
The Center for the Molecular Biology of RNA, University of California at Santa Cruz, Santa Cruz, CA, United States

André F. Seixas
Instituto de Tecnologia Química e Biológica António Xavier, Universidade Nova de Lisboa, Oeiras, Portugal

Minjung Song
Department of Food Biotechnology, College of Medical and Life Science, Silla University, Seoul, South Korea

PREFACE

The series *Progress in Molecular Biology and Translational Science* is known for its monographs on scientific areas of special interest and importance. Occasionally, however, *PMBTS* publishes "eclectic volumes," which allow the series to spotlight a number of different areas of molecular biology and medicine in a single volume. This is one such volume.

Here we have selected chapters that present the state of the art in four subject areas: (1) G protein-coupled receptors; (2) the global impact of antibiotic use and improper disposal; (3) ribozymes and RNA catabolism; and (4) the clustered regularly interspaced short palindromic repeats (CRISPR)/CRISPR associated protein 9 (Cas9) system.

Chapters 1 and 2 focus on G protein-coupled receptor (GPCRs) systems, which are involved in a large variety of cellular signaling cascades. Lueptow et al. (Chapter 1) discuss the recently deorphanized GPCR GPR83, which is highly expressed in the central nervous and immune systems. They discuss the discovery of a peptide ligand, PEN, and how the GPR83/PEN system may be involved in controlling emotions (stress, anxiety, reward), feeding, and metabolism, and therefore could be a therapeutic target for the treatment of immunological, neuroendocrine, and neuropsychiatric disorders. Lymperopoulos (Chapter 2) provides an update on β-arrestin-1 and β-arrestin-2, GPCR adapter proteins that regulate GPCRs and have important, and often conflicting, roles in cardiovascular physiology. This suggests that differential control of expression levels of these proteins may have value for the therapy of heart failure, hypertension, atherosclerosis, thromboembolism, and cardiorenal disorders.

Kelly and Brooks (Chapter 3) discuss a global toxicological problem, the release of active antibiotics into the environment, that exists not because of the difficulty in developing effective therapeutics for infectious diseases, but because antibiotics have been used so successfully around the world. Improper waste water management practices have resulted in the release of huge amounts of powerful broad spectrum antibiotics, such as Cipro (Ciprofloxacin), into the environment, where they remain active and drive the development of antibiotic-resistant organisms. These organisms cause 700,000 deaths per year. Kelly and Brooks discuss this problem and what must be done in the future to evaluate its impact on public health.

Chapters 4 and 7 discuss current work on ribozymes, enzymatically active RNA sequences. Lee et al. (Chapter 4) present the current status of *trans*-splicing ribozymes, natural RNA sequence with the capacity to eliminate and simultaneously repair or reprogram target RNAs. This has made these RNA sequences attractive agents for gene therapy, which has been investigated for a number of diseases, including myotonic dystrophy, sickle cell anemia, hepatitis C, p53-linked cancers, and pancreatic cancer. Researchers also seek to better understand the chemical biology (structure and cleavage/ligation mechanisms) of ribozymes to enable structural modification leading to enhanced activity and potential in vivo use. O'Rourke and Scott (Chapter 7) report results of their work on elucidating the mechanisms of action of an interesting ribozyme, the "hammerhead" ribozyme, and discuss how this knowledge may contribute to the design of more efficient versions of both full-length and truncated sequences.

RNA expression, function, and biological half-life must be carefully regulated in cells, be they prokaryotic or eukaryotic. One regulatory mechanism involves ribonucleases, which can cleave RNAs internally (endoribonucleases) or at their termini (exoribonucleases). These enzymes assist in the maturation of new RNA transcripts, the elimination of defective RNAs (including mRNAs), and ribonucleotide recycling. In Chapter 5, dos Santos et al. provide an overview of the function, structure, regulation, and substrate specificities of two types of eukaryotic exoribonucleases, PNPase and RNaseII/RNase R. These enzymes are key players in RNA surveillance, with activities directed at both coding and noncoding RNAs.

The CRISPR/Cas9 gene editing system is revolutionizing molecular biology and medicine. The CRISPR/Cas9 system has the potential to quickly, accurately, and cheaply silence genes, remove and replace mutant sequences with normal sequences, create DNA-directed antibiotics, eliminate HIV sequences from host T cells, and improve GMO-based crop characteristics, among many of its uses. To do so, the CRISPR/Cas9 components must be delivered to target cells. However, delivery presents a number of challenges that must be overcome before routine clinical use of CRISPR/Cas9 is achieved. In Chapter 6, Chandrasekaran et al. review current delivery methods and suggest new approaches for effective delivery of CRISPR/Cas9 in molecular medicine settings.

David B. Teplow, Ph.D. Editor

CHAPTER ONE

Targeting the Recently Deorphanized Receptor GPR83 for the Treatment of Immunological, Neuroendocrine and Neuropsychiatric Disorders

Lindsay M. Lueptow[2], Lakshmi A. Devi[1], Amanda K. Fakira[2]
Department of Pharmacological Sciences, Freidman Brain Institute, Icahn School of Medicine at Mount Sinai, New York, NY, United States
[1]Corresponding author: e-mail address: lakshmi.devi@mssm.edu

Contents

1. Introduction 2
   1.1 Discovery of GPR83 3
   1.2 Discovery of proSAAS and the Signaling Peptide PEN 4
   1.3 ProSAAS Expression and Function 5
2. Expression and Significance of GPR83 in the Brain 7
   2.1 Expression of GPR83 in the Mouse Brain 7
   2.2 Differential GPR83 Expression Between Mouse, Rat and Human 10
   2.3 Regulation of GPR83 Expression in the Brain 12
   2.4 Role of GPR83 in Hypothalamic Function 12
   2.5 Role of GPR83 in Stress, Reward and Learning and Memory 13
3. Role of GPR83 in Immune Function 15
   3.1 Expression of GPR83 in Immune Cells 15
   3.2 Significance of GPR83 in Immune Function 16
4. Current Understanding of GPR83 and PEN 18
   4.1 The Deorphanization of GPR83 18
5. Conclusions 19
   5.1 Relationship Between GPR83, Stress, Reward, and Immune Function: Future Research Considerations 19
   5.2 The GPR83-PEN Neuropeptide System as a Novel Therapeutic Drug Target 20
   5.3 Summary 20
Acknowledgment 21
References 21

[2] Equal contribution.

Progress in Molecular Biology and Translational Science, Volume 159
ISSN 1877-1173
https://doi.org/10.1016/bs.pmbts.2018.07.002

Abstract

G-protein coupled receptors (GPCRs) are a superfamily of receptors responsible for initiation of a myriad of intracellular signaling cascades. Currently, GPCRs represent approximately 34% of marketed pharmaceuticals, a large portion of which have no known endogenous ligand. These orphan GPCRs represent a large pool of novel targets for drug development. Very recently, the neuropeptide PEN, derived from the proteolytic processing of the precursor proSAAS, has been identified as a selective, high-affinity endogenous ligand for the orphan receptor, GPR83. GPR83 is highly expressed in the brain, spleen and thymus, indicating that this receptor may be a target to treat neurological and immune disorders. In the brain GPR83 is expressed in regions involved in the reward pathway, stress/anxiety responses, learning and memory and metabolism. However, the cell type specific expression of GPR83 in these regions has only recently begun to be characterized. In the immune system, GPR83 expression is regulated by Foxp3 in T-regulatory cells that are involved in autoimmune responses. Moreover, in the brain this receptor is regulated by interactions with other GPCRs, such as the recently deorphanized receptor, GPR171, and other hypothalamic receptors such as MC4R and GHSR. The following review will summarize the properties of GPR83 and highlight its known and potential significance in health and disease, as well as its promise as a novel target for drug development.

1. INTRODUCTION

G-protein coupled receptors (GPCRs) are a superfamily of receptors that convert extracellular signals into intracellular responses, via activation of various G-proteins, critical for a variety of biological functions.[1] Due to the abundance and importance of these receptors, GPCRs represent 34% of pharmaceutical drug targets. Moreover, approximately 15% of GPCRs do not have a known endogenous ligand and are therefore categorized as orphan receptors. These orphan GPCRs represent a large pool of potential therapeutic targets for various disorders.

GPR83 was originally cloned in 1989 and held the status as an orphan receptor until the recent identification of its endogenous ligand, the neuroendocrine peptide PEN, which contains a Pro-Glu-Asn tripeptide sequence.[2–4] While progress in understanding the biological role of GPR83 has been slow due to the lack of a known endogenous ligand, the use of shRNA and other viral vectors, as well as GPR83 knockout mice, has allowed some understanding of the role of GPR83 in a number of cellular and behavioral processes.

The initial investigation of GPR83 primarily involved characterization of the expression pattern of GPR83 RNA in organ systems and tissues.

Several studies have reported GPR83 in discrete brain regions, such as the hypothalamus, amygdala, hippocampus and striatum, brain areas known to regulate several neurological functions including stress, anxiety, reward and metabolic function. Interestingly, the precursor for peptide PEN (named proSAAS) is also highly expressed in brain regions that express GPR83, such as the hypothalamus, hippocampus and amygdala.[5–10] GPR83 also is expressed in various immune cells, such as B cells and CD8+ killer T cells, and more specifically, FoxP3+ regulatory T cells, but its role in immune function is not yet fully understood.[4,11–16] Taken together, these data suggest that the PEN-GPR83 receptor system is of great interest for further study for both affective disorders and immune regulation.

1.1 Discovery of GPR83

GPR83 was originally cloned from murine thymoma WEHI-7TG cells.[4,11,12] At that time researchers were searching for a glucocorticoid- and/or cAMP-inducible gene that mediated lymphocyte cytolysis, in the hope of better understanding the effects of glucocorticoids on T lymphocyte function and cell death. In doing so, they identified clone 4.2, later known as GPR83, as being up-regulated by both the synthetic glucocorticoid dexamethasone, as well as forskolin, an adenylyl cyclase activator.[4] Because of these findings, GPR83 was named glucocorticoid-induced receptor. However, in the thymoma cells, the time course for GPR83 up-regulation takes around 4 h of exposure to dexamethasone, and researchers have been unable to find any glucocorticoid response elements in the promoter region of the receptor, suggesting that the effect of glucocorticoids on GPR83 expression could be via indirect pathways.[11,12,17,18]

In 1991, the full sequence of GPR83 was published.[12] Researchers noted a few interesting aspects of the receptor, including the expression of introns in the coding region, which is somewhat unusual for GPCRs. Up to four different GPR83 transcripts have been described, and therefore, likely arise via alternative splicing of these introns.[11,18,19] Additionally, in two of the transcripts, they found differences in the second cytoplasmic loop, which is often thought to play a role in G-protein-coupling, which suggests that the receptor may be able to differentially couple to different signaling pathways via the structural variation in this region.[12] Interestingly, recent studies by the Devi group demonstrate that GPR83 differentially couples to G-proteins, depending on the brain region. They showed that GPR83 signals through the inhibitory $G\alpha_i$ G-protein in the hippocampus, but $G\alpha_q$ in the hypothalamus.[3]

While up to four splice variants have been described in mice, the only functional variant described in human brain also happens to be the most highly expressed in the mouse brain, suggesting the use of mouse tissue to study the function of the receptor is appropriate.[12,18,20] Furthermore, it is unclear if two of the rodent variants are functional, as they lack one or more transmembrane domains. Human GPR83 shares 89.5% sequence homology with mice, and 88% homology with rats, while mice and rats share 97% sequence homology.[12,20–24] The mouse GPR83 shares the highest sequence identity with the tachykinin family of receptors, with ~31–33% shared identity with the subtypes.[12] The highest degree of identity is within the transmembrane domains, with 42.4% overall identity, while the lowest homology is observed in the N-terminus, C-terminus, and third cytoplasmic loop.[12,22] However, they do share a number of potential N-linked glycosylation sites in the N-terminus, and several putative phosphorylation sites for protein kinase C and protein kinase A in the third intracellular loop and C-terminus.[22]

1.2 Discovery of proSAAS and the Signaling Peptide PEN

A critical step in the identification of an endogenous ligand for GPR83 was the discovery of a family of peptides all derived from a precursor protein termed proSAAS, so named for the amino acid sequence of one of the peptides, SAASv (Ser-Ala-Ala-Ser).[2] ProSAAS is a ~260 amino acid precursor protein that is processed into at least five peptides, SAAS, GAV, PEN, bigLEN, and littleLEN (Fig. 1).[2,10] The human form shares 84% homology with mouse/rat while the mouse and rat forms share 97% sequence homology.[5,7]

The proSAAS peptide family was discovered by neuropeptidomic studies that characterized peptides in the brain of $CPE^{fat/fat}$ mouse—a carboxypeptidase E (CPE) mouse with a known mutation that renders the enzyme inactive.[2] Since CPE is an essential enzyme in the complete processing of peptide intermediates,[25] the lack of enzymatic activity in $CPE^{fat/fat}$ results in an enrichment of these intermediate peptides, as compared to the control mice.[25,26] Using mass spectrometry, researchers were able to screen for these enriched peptides in the $CPE^{fat/fat}$ mouse brain, and thus identified the family of proSAAS peptides.[2]

Previous studies reported that mice lacking proSAAS exhibited decreased body weight,[27] whereas mice with the overexpression of proSAAS exhibited diabetes and obesity.[28] Given the potential importance of this novel peptide system, our group sought to identify the receptors

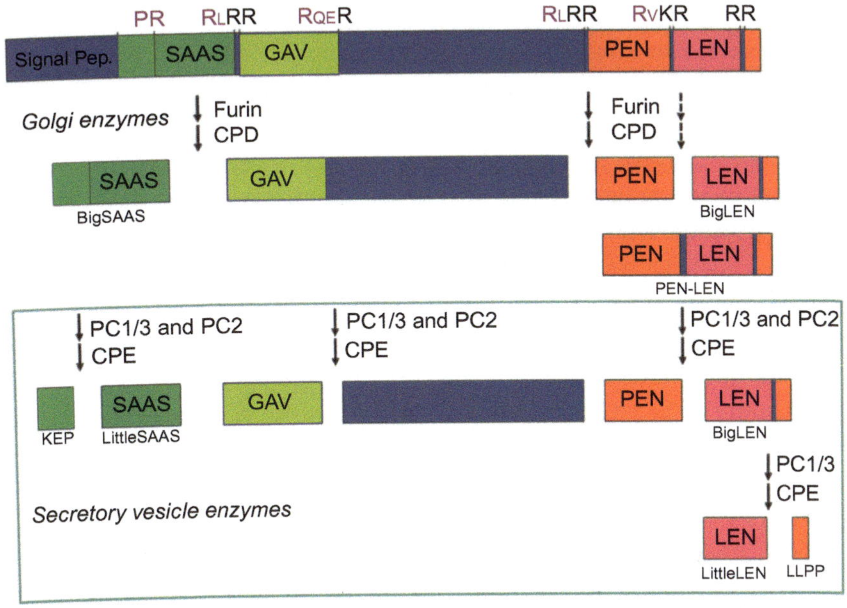

Fig. 1 Proteolytic processing of proSAAS into known signaling peptides. The precursor protein and subsequent processed peptides are indicated by color. The top row indicates the amino acid cleavage sites. The indicated residues are removed by carboxypeptidase (CPE or CPD). Below the precursor protein, the arrows indicate sites that are cleaved by furin/CPD (dashed arrow indicates potential but less efficient cleavage from furin). The box indicates enzymatic activity that occurs within the secretory vesicles. Here bigSAAS is cleaved into KEP and littleSAAS, the middle portion into GAV, and the final portion is cleaved into PEN and bigLEN, which is further processed into littleLEN and LLPP. Primary conversion of these peptides is via PC1/3 and CPE. *From Wardman JH, Fricker LD. ProSAAS-derived peptides are differentially processed and sorted in mouse brain and AtT-20 cells.* PLoS One. *2014;9:e104232, open source CCL.*

through which they might be signaling. Because of the strong expression of proSAAS in hypothalamic tissue, we started to narrow down the list of potential receptors by cross checking orphan receptors that are expressed in the hypothalamus with those expressed in specific cell lines that also show strong binding and signaling of the proSAAS peptide.[29] After finding that bigLEN binds and activates the orphan receptor GPR171, we used a similar strategy to deorphanize GPR171 orphan receptor GPR83 as the receptor that binds PEN (see Section 4 for further details).[3]

1.3 ProSAAS Expression and Function

ProSAAS is highly expressed in numerous regions of the rodent brain, but it is especially enriched in the hypothalamus, hippocampus, amygdala and other limbic regions. Data from the Allen Brain Human Atlas suggests that this expression pattern is similar in human and rodents (Fig. 2). As a relatively new peptide family, there have been a limited number of studies looking at the role of these peptides in rodent behavior. However, preliminary studies suggest that proSAAS peptides are involved in numerous behaviors, including anxiety, as well as feeding and metabolism.[9,27,28,30,31] Specifically, proSAAS knockout mice show anxiety-like behavior in the open-field and elevated O-mazes.[27] Additionally, treatment with cocaine results in a significant down-regulation of proSAAS peptides in mouse nucleus accumbens (NAc) and ventral tegmental area, two brain regions highly related to reward. proSAAS knockouts also are less susceptible to cocaine and amphetamine-induced increases in locomotor activity and behavioral sensitization.[30] ProSAAS knockout mice show a slight decrease in body weight,

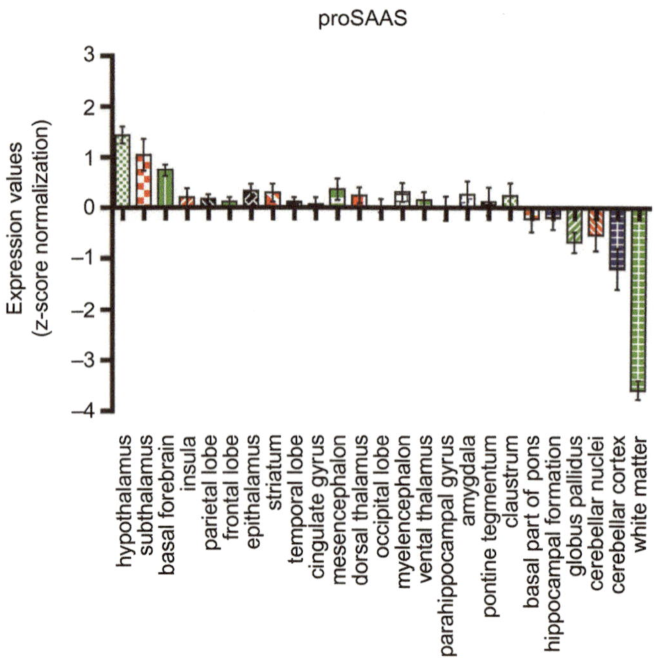

Fig. 2 The Allen Human Brain Atlas microarray data demonstrating differential expression of proSAAS across several brain regions. Expression data for six adults were obtained using two probes for PCSKIN (proSAAS). Data for each probe was averaged within each patient and then the average across patients was plotted. Data are represented as normalized z-scores and plotted from highest expression to lowest. Technical information about microarray setup and analysis are available at www.alleninsttute.org. The entire data set is available at http://human.brain-map.org.

while a global overexpression of proSAAS results in increases in body weight after 10 weeks of age.[27,28] Finally, injection of PEN antibody into the mouse third ventricle results in up to a 50% decrease in food intake that persists up to 14 h.[9]

2. EXPRESSION AND SIGNIFICANCE OF GPR83 IN THE BRAIN

2.1 Expression of GPR83 in the Mouse Brain

Initial expression studies examining GPR83 in the mouse brain found high mRNA expression in numerous regions.[4] Subsequently, several other studies characterized the region-specific expression of this receptor in mouse, rat and human.[21,23,24]

In the mouse brain, Pesini et al.[23] noted the strongest expression of GPR83 in regions that regulate reward processing, fear responses and metabolic function. These results are in agreement with data from the Allen Brain Developing Mouse Brain Atlas (Fig. 3). Of note is the expression of GPR83 in the ventral striatum, including the nucleus accumbens and olfactory tubercle (Fig. 3A and E). While both the NAc and olfactory tubercle are involved in reward, the NAc is the central region of the brain where reward signals are processed. Dopamine, the neurotransmitter that is most important in reward signaling, is released from dopaminergic neurons originating in the ventral tegmental area. Dopamine then activates either D1-type or D2-type dopamine receptors expressed on medium spiny neurons, the principal neurons in the NAc. The remaining neurons are either GABAergic or cholinergic interneurons. The neurons that specifically express GPR83 are currently unknown.

GPR83 is also highly expressed in the basolateral amygdala (BLA) and subiculum, which both send dense glutamatergic projections to the NAc[32,33] (Fig. 3C, D, I, and O). These projections are responsible for encoding reward valence[34] and regulating drug reinstatement.[35] In fact, direct activation of BLA to NAc projections is sufficient to induce reward seeking behaviors.[32] The subiculum, the output region of the ventral hippocampus, sends contextual and spatial information to the NAc to produce drug-context associations.[36] Projections from the subiculum to the NAc are selectively potentiated in response to cocaine conditioning, mediate cocaine-induced locomotion and regulate cocaine seeking behavior.[33,37,38] Expression of GPR83 in these brain regions suggests the receptor could function as a modulator of reward valence and learned associations between the reward and contextual cues.

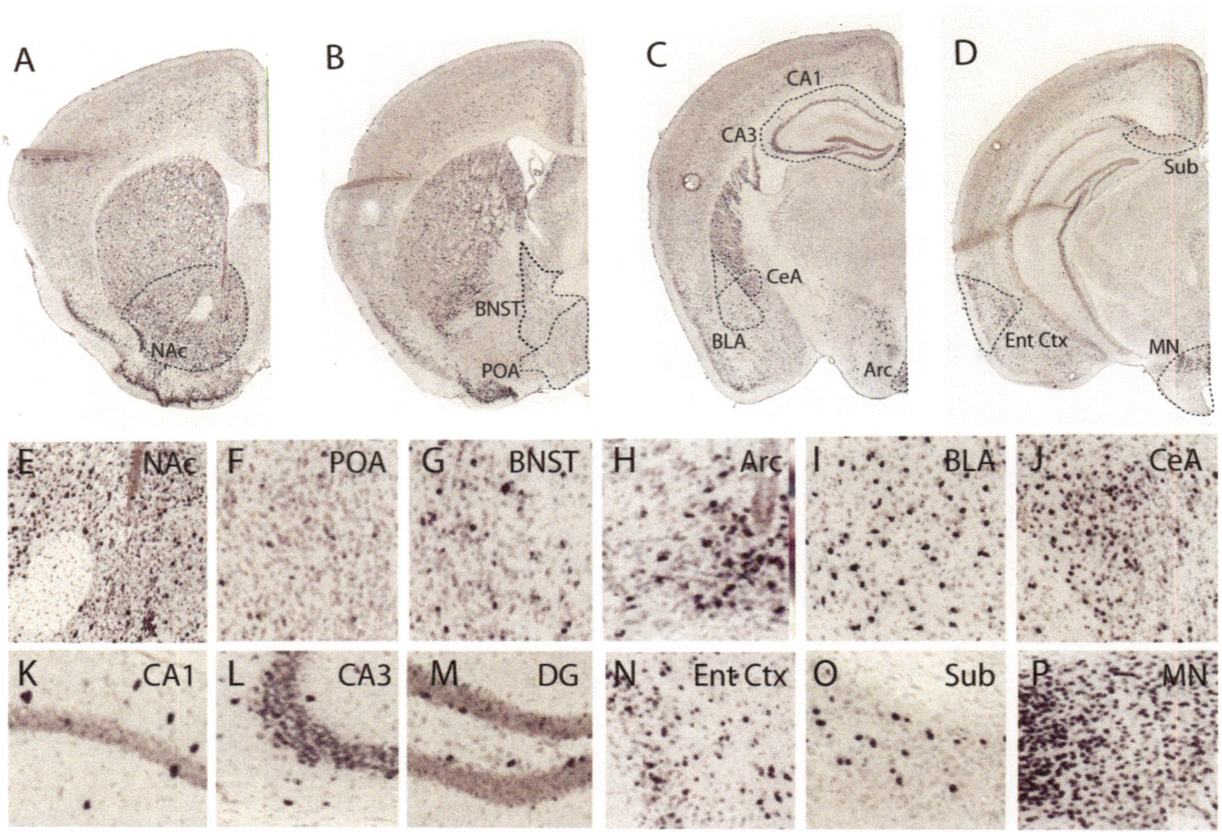

Fig. 3 GPR83 mRNA expression from the Allen Brain Developing Mouse Atlas. Coronal mouse brain sections were probed with GPR83 in situ hybridization probes. Images were obtained at http://mouse.brain-map.org/experiment/show?id=72338696. (A–D) Images from whole mouse brains probed for GPR83. Specific sub regions are outlined in black and representative larger images are displayed in (E–P). Nac, nucleus accumbens; POA, preoptic area; BNST, bed nucleus of the stria terminalis; Arc, arcuate nucleus; BLA, basolateral amygdala; CeA, central nucleus of the amygdala; CA1, Cornu Ammonis 1 (hippocampus); CA3, Cornu Ammonis 3 (hippocampus); DG, dentate gyrus; Ent Ctx, entorhinal cortex; Sub, subiculum and MN, mammillary nuclei. *Image Credit: Allen Institute.*

GPR83 has ~fivefold higher expression in the lateral habenula (whose activation produces aversion) as compared to the expression in the medial habenula, whose activation produces positive reinforcement ("reward").[39,40] Given that the regulation of the reward system by the habenula is mediated by direct modulation of dopaminergic neuron activity, the differential expression of GPR83 in this brain region places the receptor in a position to directly regulate reward signals.

Processing of emotional memory and fear responses is another function of the amygdala. Incoming sensory information is received in the lateral amygdala, which sends projections to the BLA, a region of the amygdala that has dense expression of GPR83 (Fig. 3C and I). The BLA contains predominantly glutamatergic neurons, in addition to parvalbumin and somatostatin expressing GABAergic interneurons.[41,42] The central nucleus of the amygdala (CeA), which receives inputs from the BLA, also has dense expression of GPR83 (Fig. 2C and J). In contrast to the BLA, the CeA contains GABAergic MSNs, similar to the striatum, some of which express somatostatin.[41,42] Activation of specific neuronal subtypes that form the circuit between the BLA-CeA can either produce a fear or an anxiolytic response.[42] An essential key to understanding the role of GPR83 in these processes is to characterize the neuronal subtypes in which GPR83 is expressed.

In the hypothalamus, GPR83 is expressed in the arcuate nucleus and the preoptic area[43–45] (Fig. 3B, C, F, and H). GPR83 is localized to agouti related peptide (AgRP)- and ghrelin receptor (Ghsr)-expressing neurons. AgRP neurons, which co-release NPY and GABA, sense peripheral hormonal signals such as insulin, leptin and ghrelin which act to increase feeding behaviors.[46–49] In contrast, activation of proopiomelanocortin (POMC) neurons, also localized in the arcuate nucleus, decreases feeding behaviors. In fact, studies demonstrate that GABA release from AgRP neurons increases inhibitory currents on POMC neurons and regulates energy expenditure.[50] The localization of GPR83 in this brain region implies a role for this receptor in modulating feeding behavior. In addition to its expression in the arcuate nucleus, GPR83 is expressed in warm-sensitive neurons, which are neurons that increase their firing in response to small (1–3 °C) changes in local temperature, of the preoptic area that are involved in regulating core body temperature and fever responses (Fig. 3B and F).[44] GPR83 expression in AgRP and warm-sensitive neurons suggests a significant role in regulating metabolism, energy expenditure and regulation of core body temperature, which will be discussed later in this review.[43,45]

The expression of GPR83 in the hippocampus and mammillary bodies suggests a role in regulating memory-related behaviors (Fig. 3C, D, K–M, and P). The hippocampus encodes contextual and spatial memories via the excitatory trisynaptic pathway. Here incoming information enters the hippocampus via projections from the entorhinal cortex, which synapses onto the dendrites of granule cells in the dentate gyrus. These neurons send projections to the dendrites of CA3 pyramidal neurons, which then synapse onto the dendrites of CA1 pyramidal neurons. The cellular basis of learning and memory is thought to be encoded at the synapse between CA3 and CA1 pyramidal neurons. In fact, in the mouse, GPR83 expression is strongest in the CA3[1] and dentate gyrus (our observations), suggesting that GPR83 could be involved in regulating the encoding of memories.

Another indication that GPR83 may play a role in the regulation of spatial and contextual memories is based on its high expression levels in the mammillary bodies (Fig. 3D and P).[23] One main output from the hippocampus is through the subiculum, as mentioned previously. The mammillary nuclei are a relay between subiculum outputs from the hippocampus to thalamic nuclei and have also been implicated in memory formation.[51] There is some evidence to suggest that GPR83 does in fact modulate spatial memories.[52]

2.2 Differential GPR83 Expression Between Mouse, Rat and Human

As compared to mouse GPR83, rat and human share 99% and 89.5% sequence identities respectively,[20,24,53] suggesting that the function of GPR83 is similar across species as well. The expression pattern of GPR83 among mouse, rat, and human are generally similar with some differences in the levels of expression (Fig. 4).[23,24] One difference is that GPR83 expression in the mouse hippocampus is highest in the CA3 region, while in rat, expression was reported to be equal throughout all regions of CA areas and dentate gyrus.[22,24] In contrast, in human hippocampus, GPR83 expression is strong in the dentate gyrus and CA3 regions, similar to the findings in mice.[21] Another difference between mouse and rat is GPR83 expression in the striatum. In mouse brain, GPR83 is found throughout the striatum, including core and shell regions of the nucleus accumbens with scattered expression throughout the caudate putamen.[22,23] In rat brain, GPR83 expression in the striatum is lower with conflicting reports on whether expression is localized to the core or shell regions of the nucleus accumbens.[22,24] Data from human suggests that while GPR83 is expressed in this region it is at a moderate level (Fig. 4). Overall in the human,

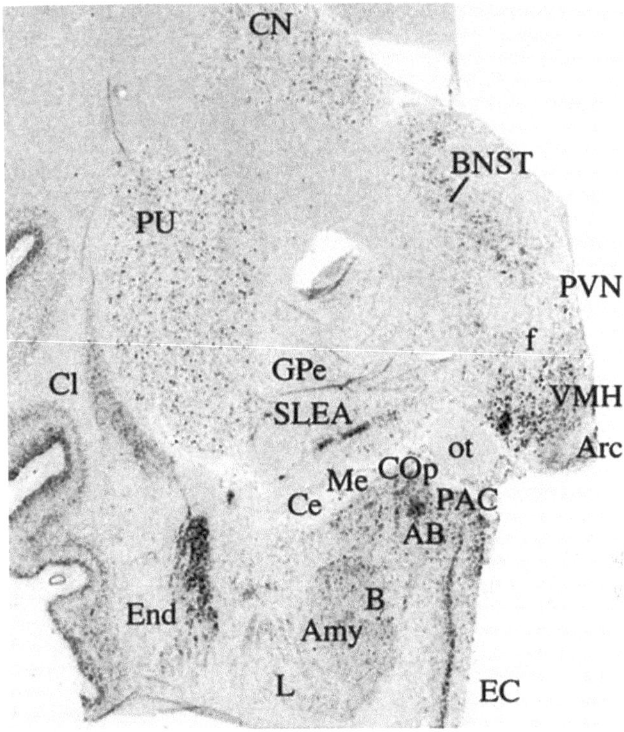

Fig. 4 GPR83 expression in the human brain in coronal 20-mm-thick sections of hemisphere showing distribution of GPR83 (JP05) mRNA using antisense riboprobe. AB, accessory basal amygdala nucleus; Amy, amygdala; Arc, arcuate; B, basal amygdaloid nucleus; BNST, bed nucleus stria terminalis; Ce, central amygdala; CN, caudate nucleus; Cop, cortical amygdaloid nucleus, posterior; Cl, claustrum; EC, entorhinal cortex; End, endopiriform cortex; f, fornix; GPe, external segment of the globus pallidus; L, lateral amygdaloid nucleus; Me, medial amygdala; ot, optic tract; PAC, periamygdaloid cortex; Pu, putamen; PVN, paraventricular hypothalamic nucleus; SLEA, sublenicular extended amygdala; VMH, ventromedial hypothalamic nucleus. *Adapted fig. 2B from Brezillion S, et al. Distribution of an orphan G-protein coupled receptor (JP05) mRNA in the human brain.* Brain Res. *2001; 921(1–2):21–30, by permission of Elsevier.*

GPR83 expression is strongest in hypothalamic, hippocampal and amygdaloid regions with moderate expression in striatal regions (Fig. 4).

One striking difference is that GPR83 is expressed in the cerebellum in the rat but not the mouse. Moreover, in the human cerebellum the strongest expression of GPR83 is in the cerebellar granular cell layer,[21] suggesting that GPR83 could play a role in regulating motor learning and coordination in humans. Overall, the pattern of expression of GPR83 suggests that the receptor plays a role in emotion, learning, reward processing and metabolic functions.

2.3 Regulation of GPR83 Expression in the Brain

The regulation of GPR83 expression in specific brain regions may be altered by neuroadaptations that occur due to stress, reward pathway activation, neuroendocrine responses and learning and memory. The first reports on GPR83 identified that the glucocorticoid agonist dexamethasone regulates the expression of the receptor in vitro.[4] Since glucocorticoids are released and bind to glucocorticoid receptors in response to activation of the hypothalamic–pituitary–adrenal (HPA) axis, this implies that GPR83 plays a role in stress responses. In addition, in vivo studies demonstrated that oral dexamethasone treatment produces a decrease in GPR83 expression in a variety of brain regions, including striatum, hippocampus and hypothalamus.[54,55] Similarly, both diet-induced obesity and fasting produces decreases in GPR83 expression in the hypothalamus,[43] suggesting that GPR83 expression is compensating for changes in metabolic function. In terms of drug reward, it has been shown that behavioral sensitization to amphetamine increases GPR83 expression in the prefrontal cortex for up to 7 days after withdrawal.[22] A recent study demonstrated that GPR83 may play a role at the intersection of reward and learning as it was demonstrated that GPR83 expression is up-regulated in cultured hippocampal astrocytes following exposure to dopamine.[56] Finally, depletion of the long chain fatty acid docosahexaenoic acid, which results in a decrease in the rate of learning of an olfactory discrimination task, decreases GPR83 expression in the olfactory bulb.[57] These studies begin to highlight the variety of neurological functions wherein GPR83 may be playing a critical role, and which need to be explored further using mouse physiology and behavior.

2.4 Role of GPR83 in Hypothalamic Function

The first reports of the in vivo impact of centrally expressed GPR83 demonstrates an important role for GPR83 in metabolic function. A shRNA viral mediated knockdown of GPR83 in the preoptic area of the anterior hypothalamus (POA) decreased core body temperature and increased body weight in mice with no changes in food intake.[45] As mentioned earlier, GPR83 is expressed in warm-sensitive neurons in the POA (Fig. 3B and F), which are responsible for maintaining core body temperature and fever responses. These neurons are GABAergic since they express mRNA for GAD 1, a GABA synthesizing enzyme, in addition to somatostatin and dynorphin peptides among others.[44] Warm-sensitive neurons of the POA project to the rostral raphe pallidus. The POA-rostral raphe projection is involved in fever initiation since immune signals that block the POA mediated inhibition of the rostral raphe result in a rise in body temperature.[58]

Further support for the role of GPR83 in hypothalamic function was demonstrated using GPR83 transgenic mice.[43] Under normal conditions GPR83 KO mice did not exhibit changes in body weight, food intake, glucose tolerance or insulin sensitivity. However, these mice exhibited a decrease in fat mass at 18 weeks of age.[43] A significant role for GPR83 in metabolic function was uncovered when GPR83 WT and KO mice were treated with ghrelin. Peripheral ghrelin treatment increases food intake in wild type mice, an effect that is exacerbated in GPR83 KO mice. Additionally, chronic central treatment with ghrelin produces significant enhancement of body weight, food intake and fat mass in GPR83 KO mice compared to WT mice. Moreover, GPR83 KO mice had increases in plasma leptin and decreases in free fatty acid levels. Given that GPR83 expression in the arcuate nucleus of the hypothalamus is localized to neurons that co-express the ghrelin receptor, these data strongly support the notion that GPR83 is influencing ghrelin receptor activity to regulate feeding behavior.

Interestingly, a study examining the effect of a high fat diet in GPR83 WT and KO mice found that the loss of GPR83 protected mice from diet-induced obesity, since they exhibited decreased weight gain and fat mass and an improved glucose tolerance compared to WT controls on the same high fat diet.[43] Moreover, although the GPR83 KO mice increased food intake compared to WTs, they did not gain as much weight as WTs on the same diet. This effect was due to increases in energy expenditure in GPR83 KO mice that were not associated with increases in locomotor activity. Increases in AgRP, NPY, Hcrt and Ghsr1a expression in the hypothalamus of GPR83 KO mice led the authors to suggest that GPR83 interacts with additional signaling mechanisms to regulate energy metabolism.[43]

2.5 Role of GPR83 in Stress, Reward and Learning and Memory

The expression of GPR83 in the amygdala, nucleus accumbens, prefrontal cortex and hippocampus (Figs. 3 and 4) suggests that this receptor plays a role in stress responses, the reward pathway and learning and memory. This is supported by studies by Vollmer et al.[52] using GPR83 KO mice. Using an acute restraint stress model, the authors demonstrated that GPR83 KO mice are protected against stress-induced anxiety, as compared to WT mice. However, GPR83 KO mice did not show any differences in serum corticosterone levels in response to restraint stress, suggesting that the mechanism of resilience is not directly related to the activation of the HPA axis. One possibility is that cell type expression of GPR83 in a brain region such as the amygdala is producing these anxiolytic effects.

Previous studies reported that repeated treatment with amphetamine to produce behavioral sensitization increases GPR83 expression in the prefrontal cortex.[22] This suggests that GPR83 expression is involved in the neuroadaptations that occur in response to drugs of abuse, which result in drug dependence. To investigate GPR83's role in the reward system, GPR83 WT and KO mice were tested in a two-bottle choice sucrose drinking paradigm where mice are presented with one bottle containing water and the other containing increasing percentages of sucrose.[52] This test is used to measure anhedonia (inability to feel pleasure), which can by induced by chronic stressors that result in depression. Neuroadaptations that result in anhedonia typically occur in the reward system of drug dependent individuals. In this assay, GPR83 KO mice had an increase preference for lower doses of 1% sucrose and not 3%, 10% or 30% sucrose.[52] Perhaps the lack of sucrose preference at higher doses could be due to a ceiling effect of the ability of sucrose to activate the reward pathway. Also, sucrose is a natural reward and it is not known whether natural reward increase GPR83 expression in a similar way as they do drugs of abuse, such as amphetamine. Overall, these studies indicate that GPR83 does play a role in reward.

The pattern of GPR83 expression in the hippocampus, prefrontal cortex and amygdala suggests that this receptor may regulate spatial and fear learning (Fig. 3).[21,23,24] Spatial learning was assessed in GPR83 WT and KO mice using the Morris water maze where mice use spatial cues to learn the location of a platform hidden beneath the surface of the water.[52] In this test, GPR83 KO mice displayed a borderline delay in the acquisition of this task and no differences in recall 24 h following training compared to WT mice. Since the Morris water maze is a hippocampal-dependent task, this suggests that GPR83 is playing a minor role in hippocampal function.[59] Contextual fear conditioning was used to determine the role of GPR83 in processing fearful stimuli. This task involves circuitry between the amygdala and prefrontal cortex and hippocampus. In GPR83 KO mice there were no differences in the acquisition, recall or extinction of fear conditioned cues suggesting that GPR83 does not play a role in these processes.[52]

One caveat of these studies is that they primarily utilized the global GPR83 KO mice instead of a region-specific knock down. In studies where the global GPR83 KO produces a biological effect in the animal, it is still unclear precisely which brain region may be responsible for producing the effect. An example is the case in which GPR83 KO alleviates anxiety produced by acute restraint stress. Given that GPR83 is highly expressed in two different subregions of the amygdala, the basolateral and central nucleus of the amygdala, as well as the bed nucleus of the stria terminalis, all of which are known to regulate anxiety and stress responses, it is not clear which region is responsible for these behavioral phenotypes. In examples where there was no effect or mild effects on spatial and fear learning with global KO of GPR83, one cannot rule out the potential for compensatory mechanisms, which may mask GPR83's role.

To date only one study has explored GPR83's direct regional effects by region specific knock-down of GPR83 in POA,[45]; a region known to regulate core body temperature and fever responses. Thus, additional studies are needed to evaluate the role of GPR83 in specific brain regions. A critical step toward completing this goal was achieved by deorphanization of GPR83.[3] By understanding the signaling pathways activated by GPR83 binding to its endogenous ligand, we can begin to explore how these signaling mechanisms affect neuronal function to produce behavioral effects.

3. ROLE OF GPR83 IN IMMUNE FUNCTION

3.1 Expression of GPR83 in Immune Cells

Although GPR83 was originally cloned in a murine thymoma cell line,[4] little was known regarding its expression on specific immune cells or its role in immune function. In 2006, Sugimoto and colleagues identified GPR83 as a marker for a specific pool of naturally occurring, activated T-regulatory cells, $CD4^+$ $CD25^+$ $Foxp3^+$ T_{regs}, which are critically important for immunological self-tolerance.[14] Deletion of either CD25 (also known as IL-2 receptor alpha chain) or the transcription factor Foxp3 (Forkhead box P3) results in a loss of natural T_{regs}, likely through the inability of naïve $CD4^+$ $CD25^-$ T_{regs} to mature, ultimately resulting in severe autoimmune disease.[60–63] Interestingly, GPR83 and Foxp3 show a very strong positive correlation in expression in the natural T_{reg} population in rodents and humans, and both are now commonly used as markers for this population of activated cells.

A second population of T-regulatory cells, sometimes referred to as suppressor cells, or T_{supp}, have also been shown to express GPR83. These cells, which are $CD8^+$ $CD25^+$ $Foxp3^+$, have a similar suppressor function as the $CD4^+$ T_{regs}, but scientists have a limited understanding of their precise biological role.[16,64] T_{supp} cells have been implicated in inflammatory diseases, specifically in human inflammatory bowel disease and asthma.[64] As with $CD4^+$ T_{regs}, following activation, $CD8^+$ T_{supp} up regulate both FoxP3 and GPR83.[16]

No direct studies have been published looking at the protein expression of GPR83 in other immune cell populations, such as B cells and NK cells. However, according to various online databases (ARCHS and BioGPS), which compile data from various RNA-seq and microarray studies, GPR83 is also expressed in a variety of immune cells including B cells, granulocytes, and neutrophils.[65,66]

3.2 Significance of GPR83 in Immune Function

Though the expression of GPR83 on natural $CD25^+$ $Foxp3^+$ T_{regs} is well established, the functional significance of this expression is still controversial. Transfection of Foxp3 leads to an increase in GPR83, and GPR83 expression is increased in mature $CD4^+$ $CD25^+$ T_{regs} that also express Foxp3, suggesting a close association between GPR83 and these mature T_{regs} (Fig. 5).[13,14,19] Hansen and colleagues showed that under inflammatory conditions in vivo, GPR83 overexpression in naïve $CD4^+$ $CD25^-$ T cells is able to induce peripheral maturation of these cells, as indicated by an increase Foxp3 expression, and confer anti-inflammatory properties.[13] However, it is important to note that when GPR83 is overexpressed on the same cells in vitro, there is no increase in Foxp3, nor are the cells able to suppress proliferation of naïve $CD4^+$ $CD25^-$ T cells. Furthermore, in vivo expression of GPR83 transfected naïve $CD4^+$ $CD25^-$ T cells is not able to induce peripheral maturation under normal or non-inflammatory conditions.[13,19] This suggests that the ligand or mode for activation of GPR83 leading to maturation of naïve T_{regs}, directly or indirectly, is only available under inflammatory conditions in vivo.

Lu et al. created a GPR83 knockout mouse to seek a better understanding of the role of the receptor in $CD4^+$ $CD25^-$ T cell maturation.[15] They found that the lack of GPR83 did not hinder T_{reg} development, nor did it alter the suppressive abilities of mature T_{regs}.[15] Additionally, Toms et al.[67] also generated a GPR83 knockout mouse, and they did not observe any differences in expression levels for naïve or mature T_{regs} in numerous different tissues, including spleen, lymph nodes, thymus, and the lamina propria of the colon.[67] In their model of colitis, lack of GPR83 did not affect the acquisition or suppression of the inflammatory disease.[67]

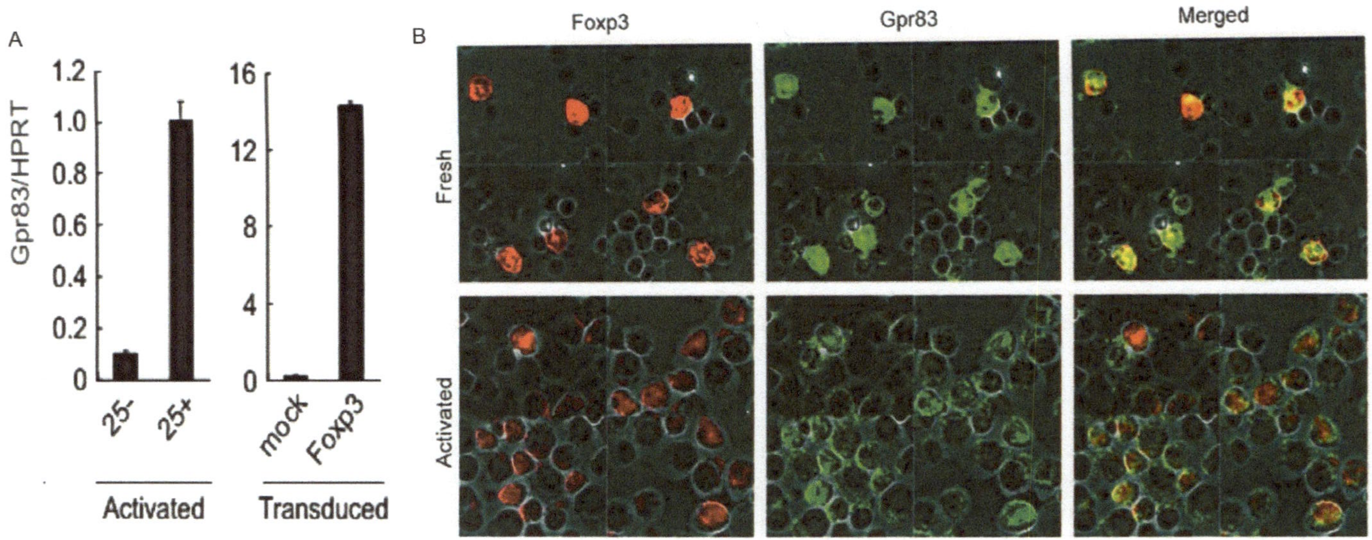

Fig. 5 Co-expression of GPR83 and Foxp3 in regulatory T-cells. (A) RT-PCR quantitative analysis of GPR83 in fresh or activated spleen and lymph node cells (left), as well as $CD25^-$ $CD4^+$ T_{regs} cells spleen and lymph node cells transduced with Foxp3- or mock- transduced (right). (B) Spleen and lymph cells were purified and freshly prepared or activated for 3 days, followed by staining with GPR83 antibodies (green) and Foxp3 antibodies (red). In fresh cells (top panel), GPR83 and FoxP3 are co-expressed in a few cells; however, following activation (bottom panel), both GPR83 and FoxP3 are greatly up-regulated in the same cell populations. *Modified figs. 4A and 5B from Sugimoto N, et al. Foxp3-dependent and -independent molecules specific for CD25+ CD4+ natural regulatory T cells revealed by DNA microarray analysis.* Int Immunol. *2006; 18(8):1197–1209, by permission of Oxford University Press.*

Taken together, no one consensus has been reached regarding the precise role of GPR83 in immune function. However, the immune landscape is quite complicated, and there are likely numerous factors and systems at play that have not been taken into account. Each study utilized different systems (i.e., skin sensitization vs intestinal inflammation) and under different conditions (i.e., overexpression of GPR83 vs depletion of GPR83), making direct comparisons and conclusions difficult to deduce. Furthermore, due to the complicated nature of immune activation, maturation, and function, there is likely significant overlap that essentially results in redundancies that are able to manage loss of function in one system. While the lack of GPR83 appears to be dispensable for proper immune function, overexpression of the receptor does suggest the ability to play a role in anti-inflammatory processes. From a therapeutic point of view, perhaps targeting a system with more of a selective, modulatory role is actually desirable, as the chance for off-target effects or undesirable side effects may be avoided. Future research into GPR83 in immune function must take these factors into consideration, and researchers may even consider a more integrated, systems-biology approach to these multi-faceted issues.

In summary, at this point, there are few definitive roles that can be attributed to GPR83 in terms of immune function. However, we do know that its expression is tightly correlated with that of Foxp3 and that its role in producing functional natural T_{regs} requires an inflammatory in vivo environment.[13,19] While it may not be a critical component of T_{reg} maturation or function, or a primary target for immune function, it may be a viable secondary target for modulatory regulation of immune responses.

4. CURRENT UNDERSTANDING OF GPR83 AND PEN

4.1 The Deorphanization of GPR83

From previous studies showing that intracerebroventricular injection of antibodies to the proSAAS derived peptides bigLEN or PEN leads to a reduction in acute feeding,[9] the receptor for these two peptides were predicted to localize in feeding centers in the hypothalamus. Hence orphan receptors enriched in the hypothalamus were screened for activation by PEN. From these studies it was found that the PEN peptide specifically activated G-protein signaling via GPR83, a receptor that was known to be expressed in the AgRP neurons that regulate feeding behavior.[3,43] Binding studies demonstrated that PEN specifically binds to brain regions that also express GPR83, including the striatum, olfactory bulb, hypothalamus and hippocampus. Moreover, these studies identified that stimulation of GPR83 by PEN activates $G\alpha_i$ mediated signaling in the hippocampus and $G\alpha_q$ signaling in the hypothalamus, indicating that GPR83 may differentially regulate downstream pathways depending on the brain region in which it is expressed. Previous studies had suggested that the neuropeptide NPY may be a ligand for GPR83 because of its high homology of GPR83 with the NPY-Y_2 receptor.[68] However Gomes et al.[3] demonstrated that NPY was not able to displace PEN at physiological concentrations. Together, these studies identified that PEN is the endogenous ligand for GPR83.

5. CONCLUSIONS

5.1 Relationship Between GPR83, Stress, Reward, and Immune Function: Future Research Considerations

GPR83 has a very interesting expression pattern, as it is highly expressed in the central nervous system (CNS) and immune systems of rodents and humans.[13,20–24,53] For decades it was assumed that these two biological systems functioned independently and that neither was able to affect or influence the other. More recently, scientists have found that not only does the immune system greatly affect the CNS but that the CNS is able to influence the peripheral immune system.[54,69–82] Some of the first indications of such interactions were in the context of stress, specifically following the activation of the HPA-axis, which results in release of corticosterone in mice (cortisol in humans).[83–87] Generally speaking, acute stress and subsequent release of corticosteroids results in temporary pro-immune or immunoprotective responses, while chronic stress can depress immune function and result in aberrant immune-pathological states.[83,85,88,89] Furthermore, chronic exposure to corticosteroids results in numerous cellular changes in the brain, and chronic stress is a strong risk factor for the development of anxiety, depression, addiction, and other psychological disorders.[80,83,86,90,91] GPR83 was discovered as being activated by the synthetic glucocorticoid dexamethasone, suggesting that GPR83 may be involved in the stress response, and indeed, knocking out GPR83 in mice leads to a stress protective phenotype.[52] Further research needs to be conducted regarding the role of the immune system in response to stress and the potential protective effects GPR83 may play centrally or peripherally. Because its neuro-immuno-endocrine expression pattern, it is likely that there is a strong interaction between GPR83, stress, and immune function that should be further explored.

5.2 The GPR83-PEN Neuropeptide System as a Novel Therapeutic Drug Target

Despite a somewhat limited understanding of the biological role of GPR83, development of small molecule compounds that can bind to this receptor will assist in the functional characterization of the receptor in a variety of psychological and immune disease states. For example, a peripherally restricted compound could target GPR83 function in only immune cells. As discussed above, knockout of GPR83 does not appear to cause gross immune deficiencies. However, GPR83 expression does influence immune response to certain inflammatory conditions in vivo, suggesting that it would be an excellent drug target to modulate immune function in autoimmune or inflammatory conditions.

In the context of the CNS alone, GPR83 knockout mice are resistant to stress-induced anxiety.[52] GPR83 compounds may be viable novel targets for the development of anxiolytics with fewer side effects than currently used compounds such as benzodiazepines and SSRI/SNRIs, which can cause drowsiness, interact with other drugs, or be abused. Furthermore, GPR83 antagonists may be useful as prophylactics in the prevention of stress-induced anxiety for people in high-stress environments that are at a higher risk for development of PTSD and other stress-related psychological issues.

GPR83 knockout mice also show resistance to diet-induced increases in obesity and diet-induced glucose intolerance, and sequestration of PEN in the hypothalamus using antibodies significantly reduces feeding, suggesting an important role for GPR83 in feeding and metabolism.[9,43] Therefore, GPR83 may also be a novel target for those struggling with eating disorders and obesity, or even diet-induced diabetes.

To date there has been one study identifying GPR83 as a potential factor in the development of obstructive sleep apnea in human patients.[92] Here the authors found the genetic loci, which contain the GPR83 gene, as a risk factor for this disorder. These studies identify that GPR83 is expressed in brain regions related to this disorder and that GPR83's expression on T_{regs}, in the immune system, also may be related to the severity of obstructive sleep apnea. Together these are the first indication that GPR83 is playing a role in the intersection of the brain and immune system.

5.3 Summary

In summary, GPR83 and its endogenous ligand, PEN, exhibit region specific expression in neuroendocrine tissues, and the receptor is highly expressed in various cells of the immune system. It has been implicated in numerous behaviors, including stress, anxiety, reward, as well as feeding and metabolism. While much more research is required to fully elucidate the role of GPR83 and PEN in all of these biological systems, this neuropeptide-receptor system has a strong potential as a novel target for future drug development. Certainly, from both a basic biology perspective, as well as for therapeutic development, compounds targeting this receptor will open a wide door for future research.

ACKNOWLEDGMENT

We would like to thank Ivone Gomes for critical reading of the article and insightful comments. This work was supported by NIH grants T32007135 (L.M.L.), DA008863 and NS026880 (to L.A.D.).

REFERENCES

1. Jonas KC, Hanyaloglu AC. Impact of G protein-coupled receptor heteromers in endocrine systems. *Mol Cell Endocrinol.* 2017;449:21–27.
2. Fricker LD, et al. Identification and characterization of proSAAS, a granin-like neuroendocrine peptide precursor that inhibits prohormone processing. *J Neurosci.* 2000;20:639–648.
3. Gomes I, et al. Identification of GPR83 as the receptor for the neuroendocrine peptide PEN. *Sci Signal.* 2016;9:1–15.
4. Harrigan MT, Baughman G, Campbell NF, Bourgeois S. Isolation and characterization of glucocorticoids- and cyclic AMP-induced genes in T lymphocytes. *Mol Cell Biol.* 1989;9:3438–3446.
5. Mzhavia N, Berman Y, Che FY, Fricker LD, Devi LA. ProSAAS processing in mouse brain and pituitary. *J Biol Chem.* 2001;276:6207–6213.
6. Sayah M, Fortenberry Y, Cameron A, Lindberg I. Tissue distribution and processing of proSAAS by proprotein convertases. *J Neurochem.* 2001;76:1833–1841.
7. Feng Y, Reznik SE, Fricker LD. Distribution of proSAAS-derived peptides in rat neuroendocrine tissues. *Neuroscience.* 2001;105:469–478.
8. Morgan DJ, et al. Embryonic gene expression and pro-protein processing of proSAAS during rodent development. *J Neurochem.* 2005;93:1454–1462.
9. Wardman JH, Berezniuk I, Di S, Tasker JG, Fricker LD. ProSAAS-derived peptides are colocalized with neuropeptide Y and function as neuropeptides in the regulation of food intake. *PLoS One.* 2011;6:e28152.
10. Wardman JH, Fricker LD. ProSAAS-derived peptides are differentially processed and sorted in mouse brain and AtT-20 cells. *PLoS One.* 2014;9:e104232.
11. Baughman G, Harrigan MT, Campbell NF, Nurrish SJ, Bourgeois S. Genes newly identified as regulated by glucocorticoids in murine thymocytes. *Mol Endocrinol.* 1991;5:637–644.
12. Harrigan MT, Campbell NF, Bourgeois S. Identification of a gene induced by glucocorticoids in murine T-cells: a potential G protein-coupled receptor. *Mol Endocrinol.* 1991;5:1331–1338.
13. Hansen W, et al. G protein-coupled receptor 83 overexpression in naive CD4+CD25- T cells leads to the induction of Foxp3+regulatory T cells in vivo. *J Immunol.* 2006;177:209–215.

14. Sugimoto N, et al. Foxp3-dependent and -independent molecules specific for CD25 +CD4+ natural regulatory T cells revealed by DNA microarray analysis. *Int Immunol.* 2006;18:1197–1209.
15. Lu L-F, Gavin MA, Rasmussen JP, Rudensky AY. G protein-coupled receptor 83 is dispensable for the development and function of regulatory T cells. *Mol Cell Biol.* 2007;27:8065–8072.
16. Fleissner D, et al. Generation and function of immunosuppressive human and murine CD8+ T cells by transforming growth factor-β and retinoic acid. *Immunology.* 2011;134:82–92.
17. Shuji Hinuma I, Junichi Sakamoto O, Masaki Hosoya I. *G Protein Coupled Receptor Proteins, Their Production and Use, US 2002/0009771 A1,* 1997:51.
18. AstraZeneca. *Glucocorticoid Induced Receptor (hgir,rp-23) Ligand Screening, WO/2001/044439,* 2000.
19. Hansen W, et al. Inflammation in vivo is modulated by GPR83 isoform-4 but not GPR83 isoform-1 expression in regulatory T cells. *Genes Immun.* 2010;11:357–361.
20. De Moerlooze L, Williamson J, Liners F, Perret J, Parmentier M. Cloning and chromosomal mapping of the mouse and human genes encoding the orphan glucocorticoid-induced receptor (GPR83). *Cytogenet Cell Genet.* 2000;90:146–150.
21. Brezillon S, Detheux M, Parmentier M, Hokfelt T, Hurd YL. Distribution of an orphan G-protein coupled receptor (JP05) mRNA in the human brain. *Brain Res.* 2001;921:21–30.
22. Wang D, et al. Cloning, expression and regulation of a glucocorticoid-induced receptor in rat brain: effect of repetitive amphetamine. *J Neurosci.* 2001;21:9027–9035.
23. Pesini P, Detheux M, Parmentier M, Hökfelt T. Distribution of a glucocorticoid-induced orphan receptor (JP05) mRNA in the central nervous system of the mouse. *Mol Brain Res.* 1998;57:281–300.
24. Sah R, et al. Expression of the glucocorticoid-induced receptor mRNA in rat brain. *Neuroscience.* 2005;133:281–292.
25. Zhang X, et al. Peptidomics of $Cpe^{fat/fat}$ mouse brain regions: implications for neuropeptide processing. *J Neurochem.* 2008;107:1596–1613.
26. Mzhavia N, et al. Processing of proSAAS in neuroendocrine cell lines. *Biochem J.* 2002;76:67–76.
27. Morgan DJ, et al. The propeptide precursor proSAAS is involved in fetal neuropeptide processing and body weight regulation. *J Neurochem.* 2010;113:1275–1284.
28. Wei SW, et al. Obesity and diabetes in transgenic mice expressing proSAAS. *J Endocrinol.* 2004;180:357–368.
29. Gomes I, et al. GPR171 is a hypothalamic G protein-coupled receptor for BigLEN, a neuropeptide involved in feeding. *Proc Natl Acad Sci USA.* 2013;110:16211–16216.
30. Berezniuk I, et al. ProSAAS-derived peptides are regulated by cocaine and are required for sensitization to the locomotor effects of cocaine. *J Neurochem.* 2017;143:268–281.
31. Wardman JH, et al. Analysis of peptides in prohormone convertase 1/3 null mouse brain using quantitative peptidomics. *J Neurochem.* 2010;114:215–225.
32. Stuber GD, et al. Excitatory transmission from the amygdala to nucleus accumbens facilitates reward seeking. *Nature.* 2011;475:377–382.
33. Britt JP, et al. Synaptic and behavioral profile of multiple glutamatergic inputs to the nucleus accumbens. *Neuron.* 2012;76:790–803.
34. Murray EA. The amygdala, reward and emotion. *Trends Cogn Sci.* 2007;11:489–497.
35. Fuchs RA, et al. The role of the dorsomedial prefrontal cortex, basolateral amygdala, and dorsal hippocampus in contextual reinstatement of cocaine seeking in rats. *Neuropsychopharmacology.* 2005;30:296–309.
36. Sesack SR, Grace AA. Cortico-basal ganglia reward network: microcircuitry. *Neuropsychopharmacology.* 2010;35:27–47.

37. Vorel SR, Liu X, Hayes RJ, Spector JA, Gardner EL. Relapse to cocaine-seeking after hippocampal theta burst stimulation. *Science*. 2001;292:1175–1178.
38. Sun W, Rebec GV. Lidocaine inactivation of ventral subiculum attenuates cocaine-seeking behavior in rats. *J Neurosci*. 2003;23:10258–10264.
39. Wagner F, Bernard R, Derst C, French L, Veh RW. Microarray analysis of transcripts with elevated expressions in the rat medial or lateral habenula suggest fast GABAergic excitation in the medial habenula and habenular involvement in the regulation of feeding and energy balance. *Brain Struct Funct*. 2016;221:4663–4689.
40. Fakhoury M. The dorsal diencephalic conduction system in reward processing: spotlight on the anatomy and functions of the habenular complex. *Behav Brain Res*. 2018;348:115–126.
41. Krabbe S, Gründemann J, Lüthi A. Amygdala inhibitory circuits regulate associative fear conditioning. *Biol Psychiatry*. 2017;83:800–809.
42. Janak PH, Tye KM. From circuits to behaviour in the amygdala. *Nature*. 2015;517:284–292.
43. Müller TD, et al. The orphan receptor Gpr83 regulates systemic energy metabolism via ghrelin-dependent and ghrelin-independent mechanisms. *Nat Commun*. 2013;4:1968.
44. Eberwine J, Bartfai T. Single cell transcriptomics of hypothalamic warm sensitive neurons that control core body temperature and fever response: signaling asymmetry and an extension of chemical neuroanatomy. *Pharmacol Ther*. 2011;129:241–259.
45. Dubins JS, et al. Downregulation of GPR83 in the hypothalamic preoptic area reduces core body temperature and elevates circulating levels of adiponectin. *Metabolism*. 2012;61:1486–1493.
46. Gropp E, et al. Agouti-related peptide-expressing neurons are mandatory for feeding. *Nat Neurosci*. 2005;8:1289–1291.
47. Schwartz MW, et al. Inhibition of hypothalamic neuropeptide Y gene expression by insulin. *Endocrinology*. 1992;130:3608–3616.
48. Schwartz MW, Seeley RJ, Campfield LA, Burn P, Baskin DG. Identification of targets of leptin action in rat hypothalamus. *J Clin Invest*. 1996;98:1101–1106.
49. Chen HY, et al. Orexigenic action of peripheral ghrelin is mediated by neuropeptide Y and agouti-related protein. *Endocrinology*. 2004;145:2607–2612.
50. Tong Q, Ye CP, Jones JE, Elmquist JK, Lowell BB. Synaptic release of GABA by AgRP neurons is required for normal regulation of energy balance. *Nat Neurosci*. 2008;11:998–1000.
51. Vann SD, Aggleton JP. The mammillary bodies: two memory systems in one? *Nat Rev Neurosci*. 2004;5:35–44.
52. Vollmer L, et al. Attenuated stress-evoked anxiety, increased sucrose preference and delayed spatial learning in glucocorticoid-induced receptor-deficient mice. *Genes Brain Behav*. 2013;12:241–249.
53. Parker R, et al. Y-receptor-like genes GPR72 and GPR73: molecular cloning, genomic organisation and assignment to human chromosome 11q21.1 and 2p14 and mouse chromosome 9 and 6. *Biochim Biophys Acta—Gene Struct Expr*. 2000;1491:369–375.
54. Souza GR, et al. Individual differences in the peripheral immune system promote resilience versus susceptibility to social stress. *Proc Natl Acad Sci USA*. 2014;111:18799.
55. Adams F, Grassie M, Shahid M, Hill DR, Henry B. Acute oral dexamethasone administration reduces levels of orphan GPCR glucocorticoid-induced receptor (GIR) mRNA in rodent brain: potential role in HPA-axis function. *Brain Res Mol Brain Res*. 2003;117:39–46.
56. Galloway A, et al. Dopamine triggers CTCF-dependent morphological and genomic remodeling of astrocytes. *J Neurosci*. 2018;10:3349–17.
57. Hichami A, et al. Olfactory discrimination ability and brain expression of c-fos, Gir and Glut1 mRNA are altered in n-3 fatty acid-depleted rats. *Behav Brain Res*. 2007;184:1–10.

58. Nakamura K, et al. The rostral raphe pallidus nucleus mediates pyrogenic transmission from the preoptic area. *J Neurosci.* 2002;22:4600–4610.
59. Morris RGM, Garrud P, Rawlins JNP, O'Keefe J. Place navigation impaired in rats with hippocampal lesions. *Nature.* 1982;297:681–683.
60. Sakaguchi S, Sakaguchi N, Asano M, Itoh M, Toda M. Immunologic self-tolerance maintained by activated T cells expressing IL-2 receptor alpha-chains (CD25). Breakdown of a single mechanism of self-tolerance causes various autoimmune diseases. *J Immunol.* 1995;155:1151–1164.
61. Asano M, Toda M, Sakaguchi N, Sakaguchi S. Autoimmune disease as a consequence of developmental abnormality of a T cell subpopulation. *J Exp Med.* 1996;184:387–396.
62. Bennett CL, et al. The immune dysregulation, polyendocrinopathy, enteropathy, X-linked syndrome (IPEX) is caused by mutations of FOXP3. *Nat Genet.* 2001;27: 20–21.
63. Brunkow ME, et al. Disruption of a new forkhead/winged-helix protein, scurfin, results in the fatal lymphoproliferative disorder of the scurfy mouse. *Nat Genet.* 2001;27:68–73.
64. Churlaud G, et al. Human and mouse CD8+CD25+FOXP3+ regulatory T cells at steady state and during interleukin-2 therapy. *Front Immunol.* 2015;6(171).
65. Lattin JE, et al. Expression analysis of G protein-coupled receptors in mouse macrophages. *Immunome Res.* 2008;4:5.
66. Lachmann A, et al. Massive mining of publicly available RNA-seq data from human and mouse. *Nat Commun.* 2018;9:1366.
67. Toms C, et al. Gpr83 expression is not required for the maintenance of intestinal immune homeostasis and regulation of T-cell-dependent colitis. *Immunology.* 2008;125:302–312.
68. Sah R, et al. Interaction of NPY compounds with the rat glucocorticoid-induced receptor (GIR) reveals similarity to the NPY-Y2 receptor. *Peptides.* 2007;28:302–309.
69. Szpunar MJ, Belcher EK, Dawes RP, Madden KS. Sympathetic innervation, norepinephrine content, and norepinephrine turnover in orthotopic and spontaneous models of breast cancer. *Brain Behav Immun.* 2016;53:223–233.
70. Ogłodek E, Szota A, Just M, Moś D, Araszkiewicz A. The role of the neuroendocrine and immune systems in the pathogenesis of depression. *Pharmacol Rep.* 2014;66:776–781.
71. ThyagaRajan S, Priyanka HP. Bidirectional communication between the neuroendocrine system and the immune system: relevance to health and diseases. *Ann Neurosci.* 2012;19:40–46.
72. Jouda J, et al. T cells affect central and peripheral noradrenergic mechanisms and neurotrophin concentration in the spleen and hypothalamus. *Ann N Y Acad Sci.* 2012;1261:18–25.
73. Roggero E, Besedovsky HO, Del Rey A. The role of the sympathetic nervous system in the thymus in health and disease. *Neuroimmunomodulation.* 2011;18:339–349.
74. Song J, Abraham SN. Innate and adaptive immune responses in the CNS. *Eur J Clin Invest.* 2008;38(suppl 2):21–28.
75. Bhattacharya A, Derecki NC, Lovenberg TW, Drevets WC. Role of neuro-immunological factors in the pathophysiology of mood disorders. *Psychopharmacology (Berl).* 2016;233:1623–1636.
76. Ordovas-Montanes J, et al. The regulation of immunological processes by peripheral neurons in homeostasis and disease. *Trends Immunol.* 2015;36:578–604.
77. Hodes GE, Kana V, Menard C, Merad M, Russo SJ. Neuroimmune mechanisms of depression. *Nat Neurosci.* 2015;18:1386–1393.
78. Stewart AM, et al. Cytokine and endocrine parameters in mouse chronic social defeat: implications for translational 'cross-domain' modeling of stress-related brain disorders. *Behav Brain Res.* 2015;276:84–91.

79. Padro CJ, Sanders VM. Neuroendocrine regulation of inflammation. *Semin Immunol.* 2014;26:357–368.
80. Guan SZ, et al. Chronic unpredictable mild stress impairs erythrocyte immune function and changes T-lymphocyte subsets in a rat model of stress-induced depression. *Environ Toxicol Pharmacol.* 2014;37:414–422.
81. Kelley KW, McCusker RH. Getting nervous about immunity. *Semin Immunol.* 2014;26:389–393.
82. Bellinger DL, Lorton D. Autonomic regulation of cellular immune function. *Auton Neurosci.* 2014;182:15–41.
83. Elenkov IJ, Chrousos GP. Stress, cytokine patterns and susceptibility to disease. *Baillieres Best Pract Res Clin Endocrinol Metab.* 1999;13:583–595.
84. Elenkov IJ, Chrousos GP. Stress hormones, Th1/Th2 patterns, pro/anti-inflammatory cytokines and susceptibility to disease. *Trends Endocrinol Metab.* 1999;10:359–368.
85. Dhabhar FS. Stress-induced augmentation of immune function—the role of stress hormones, leukocyte trafficking, and cytokines. *Ann N Y Acad Sci.* 2003;992:205–217.
86. Tynan RJ, et al. Chronic stress alters the density and morphology of microglia in a subset of stress-responsive brain regions. *Brain Behav Immun.* 2010;24:1058–1068.
87. Heidt T, et al. Chronic variable stress activates hematopoietic stem cells. *Nat Med.* 2014;20:754–758.
88. Dhabhar FS. Enhancing versus suppressive effects of stress on immune function: implications for immunoprotection and immunopathology enhancing vs. suppressive effects of stress on immune function. *Neuroimmunomodulation.* 2009;16:300–317.
89. Cohen S, et al. Chronic stress, glucocorticoid receptor resistance, inflammation, and disease risk. *Proc Natl Acad Sci USA.* 2012;109:5995–5999.
90. Hughes MM, Connor TJ, Harkin A. Stress-related immune markers in depression: implications for treatment. *Int J Neuropsychopharmacol.* 2016;19(6):1–19.
91. Nicolaides NC, Kyratzi E, Lamprokostopoulou A, Chrousos GP, Charmandari E. Basic research concerning glucocorticoids stress, the stress system and the role of glucocorticoids. *Neuroimmunomodulation.* 2015;22:6–19.
92. Cade BE, et al. Genetic associations with obstructive sleep apnea traits in hispanic/latino americans. *Am J Respir Crit Care Med.* 2016;194:886–897.

CHAPTER TWO

Arrestins in the Cardiovascular System: An Update

Anastasios Lymperopoulos[1]
Laboratory for the Study of Neurohormonal Control of the Circulation, Department of Pharmaceutical Sciences, Nova Southeastern University College of Pharmacy, Fort Lauderdale, FL, United States
[1]Corresponding author: e-mail address: al806@nova.edu

Contents

1. Introduction 28
2. Cardiovascular Adrenergic Receptors and βarrestins 29
   2.1 Cardiovascular αARs and βarrestins 29
   2.2 Cardiac βARs and βarrestins 31
   2.3 Other Cardiovascular βARs and βarrestins 36
3. Cardiovascular Angiotensin II Receptors and βarrestins 37
   3.1 Cardiac AT_1Rs and βarrestins 37
   3.2 Vascular AT_1Rs and βarrestins 39
   3.3 Adrenal AT_1Rs and βarrestins 41
4. Other Cardiovascular GPCRs and βarrestins 42
   4.1 Endothelin Receptors 42
   4.2 Vasopressin Receptors 43
   4.3 Niacin Receptor (GPR109A) 43
   4.4 P2Y Receptors 44
   4.5 Protease-Activated Receptors 45
   4.6 Apelin Receptor 46
   4.7 Sphingosine-1-Phosphate 1 Receptor 46
5. Therapeutic Implications of the Functional Divergence of Cardiovascular βarrestins 47
6. Conclusions and Future Perspectives 49
References 50

Abstract

Three major functional roles for the two ubiquitous G protein-coupled receptor (GPCR) adapter proteins, βarrestin1 and -2 (also known as Arrestin2 and -3, respectively), have been described: (a) functional desensitization, i.e., G-protein uncoupling from the receptor, (b) GPCR internalization via clathrin-coated pits, and (c) formation of signalosomes. Either while bound to the GPCR or after dissociating from it, and either while trafficking inside the cell or from the plasma membrane, both βarrestins are known to mediate a large part of the G protein-independent signaling elicited by GPCRs. The latter constitute the single most commonly targeted receptor class by

Progress in Molecular Biology and Translational Science, Volume 159
ISSN 1877-1173
https://doi.org/10.1016/bs.pmbts.2018.07.003

FDA-approved cardiovascular drugs, with about one-third of all currently used in the clinic medications affecting GPCR function. After initially activating heterotrimeric G proteins, most of these receptors can switch their intracellular signaling to other, G protein-independent, pathways, which are usually βarrestin-dependent. Since cardiovascular GPCRs are at the interface between hormonal actions and regulation of the circulation, cardiovascular βarrestins, by regulating these receptors in the heart, vasculature, platelets, adrenals, and other organs/tissues, play important roles in the functional regulation of the normal healthy circulatory system as well as in its dysregulation in disease. One important aspect in the study of these promiscuous signal transducers is the fact that the two βarrestin isoforms display vast, in some cases contrasting, differences in their biological functions in several organ systems, including in the systemic circulation. In this chapter, we provide an update on the physiology and pharmacology of the two βarrestins throughout the cardiovascular system and on how current knowledge can be utilized therapeutically. We also highlight the functional diversity of the two βarrestins in the circulatory system, particularly in the heart, and its impact on the translational potential of cardiovascular βarrestin therapeutic targeting.

1. INTRODUCTION

Out of the four mammalian arrestin isoforms, only the two ubiquitous (outside the retinal system) βarrestins are expressed in the mammalian cardiovascular system with βarrestin1 protein being ~10–15-fold more abundant than βarrestin2.[1] Both βarrestin isoforms regulate all non-opsin G protein-coupled receptors (GPCRs), also known as seven transmembrane-spanning receptors (7TMRs), including those responsible for neurohormonal regulation of cardiovascular physiology.[2,3] For instance, cardiac function (contractility) is tightly controlled by the activity of β-adrenergic receptors (βARs) located in the membranes of cardiac myocytes.[4–8] Cardiac structure and morphology are regulated by angiotensin II (AngII) type 1 receptors (AT_1Rs) present (mainly) in cardiac fibroblast and endothelial cell membranes.[4,7] Heart rate is under the balance of β-adrenergic and muscarinic cholinergic receptor (mAChR) activities.[4,7] Even the production and release of the regulatory hormones per se, whether it be catecholamine and corticosteroid release by the adrenal glands or activation of the renin–angiotensin–aldosterone system (RAAS) by the juxtaglomerular apparatus of the kidneys or release of neurotransmitters by central and peripheral neurons innervating cardiovascular organs, is under tight regulation by various GPCRs.[1,4,7]

Cardiovascular GPCRs can signal either through G-proteins or βarrestins with the natural, endogenous agonist hormones activating both signal transducers at each receptor fully and equally.[1,9] Several "biased" GPCR ligands have been discovered that (relatively) selectively activate either G proteins or βarrestins.[1,9] This "bias" in terms of the activated signal transducer is always relative; never absolute. Additionally, recent experimental evidence in non-native cell systems provides unexpected twists for the concept of biased signaling. Specifically, it has been suggested that a receptor can activate both a G-protein and a βarrestin at the same time, with both proteins transducing signals from the same receptor, bound to the same ligand simultaneously.[10] Additionally, it was shown very recently that βarrestins can be catalytically activated by a variety of GPCRs, much like heterotrimeric G-proteins.[11] In other words, they can associate with, and dissociate from the receptor's core spontaneously and transiently, prior to engaging clathrin-coated structural components at the plasma membrane to initiate intracellular signals from there without entering into the cytoplasm.[11] Of course, these studies raise important questions, such as how the βarrestin gets inactivated if it does not recycle through the cytoplasm, and how agonist-dependent this type of "catalytic" βarrestin activation and signaling is, given that no receptor-βarrestin complexes are formed during this "catalytic" activation process. Nonetheless, if these findings hold true in native cell systems or in vivo, selective activation of one signal transducer over the other (G-protein vs. βarrestin) by a single GPCR might prove next to impossible, rendering the concept of biased signaling for therapeutic purposes unattainable.

Since the ARs and AT_1Rs are by far the best-studied cardiovascular GPCRs in the context of βarrestins' physiological functions, we will start our overview of cardiovascular βarrestin physiology with these two cardiovascular GPCR types, and then, we will summarize the actions of βarrestins on the rest of the cardiovascular GPCRs type-by-type. Finally, we will conclude with a discussion of the functional divergence of the two βarrestin isoforms in the cardiovascular system and what that means for cardiovascular therapy.

2. CARDIOVASCULAR ADRENERGIC RECEPTORS AND βARRESTINS

2.1 Cardiovascular αARs and βarrestins

Cardiovascular α_1ARs, consisting of three pharmacological subtypes, α_{1A}, α_{1B}, α_{1D}, are the classic adrenoceptors mediating vascular smooth muscle (VSM) contraction via coupling to the $G_{q/11}$ protein-Phospholipase C-Ca^{2+} mobilization pathway, which elevates systemic vascular resistance (blood pressure) and cardiac afterload.[12] In addition, α_1ARs are present in cardiac myocytes, wherein they mediate maladaptive cardiac hypertrophy leading to heart failure (HF) by increasing expression of β-myosin heavy chain and other maladaptive genes belonging to the so-called "fetal gene program" of cardiac adverse remodeling.[13] Carvedilol, a βAR blocker currently used with great success in chronic HF therapy, is also an α_1AR antagonist; thus antagonism of the cardiotoxic maladaptive effects of cardiac α_1ARs may underlie part of this drug's beneficial actions in chronic HF.[14] However, not all actions of cardiac α_1ARs are bad. α_1ARs can also stimulate "adaptive" or "physiological" cardiac hypertrophy with no fibrosis or other manifestations of adverse remodeling but rather preservation or even augmentation of cardiac function and stimulation of myocyte proliferation with ventricular muscle wall thickening.[15,16] However, the physiological effects of βarrestins acting on cardiovascular α_1ARs are poorly understood. α_1ARs were recently shown to hetero-oligomerize with complexes of chemokine (C-X-C motif) receptor (CXCR) 4 and atypical chemokine receptor 3 (ACKR3), i.e., ACKR3:CXCR4 hetero-dimers, in human VSM cells.[17] This is required for VSM $\alpha_{1B/D}$AR function, and activation of ACKR3 inhibits α_1AR signaling and function but not βarrestin recruitment.[17] Thus, this α_1AR hetero-oligomerization in VSM may confer novel signaling properties and differential receptor regulation compared to native α_1AR monomers. Another very recent study examined the role of βarrestins in heterologous (i.e., agonist-independent) desensitization of cardiac βARs.[18] Although βarrestins are only directly involved in homologous (agonist-dependent) desensitization, they may desensitize inactive receptors indirectly, via their activation by other GPCRs. Indeed, βarrestin2, upon its activation by the α_1AR, among other GPCRs, was shown to recruit phosphodiesterase 4D (PDE4D) to the β_2AR and desensitize it in neonatal rat ventricular myocytes (NRVMs).[18] The recruitment of PDE4D to the β_2AR effectively terminated the 3′,5′-adenosine monophosphate (cAMP) signaling induced by isoproterenol stimulation of the receptor. The mechanism appeared to involve protein kinase A (PKA)-dependent phosphorylation of the β_2AR, since mutation of the putative β_2AR PKA phosphorylation sites, inhibition of PDE4, or genetic ablation of βarrestin2 abolished this heterologous inhibitory effect.[18] However, how the α_1AR, normally a $G_{q/11}$-coupled receptor, activates PKA to desensitize the β_2AR, and whether the same mechanism applies to the much more abundant (than the β_2AR) β_1AR in cardiac myocytes[4,19] remain two open questions from this study.

Much more is known about βarrestin-dependent effects on the α_2ARs in the cardiovascular system. A polymorphic variant of the human α_{2B}AR gene (Del 301–303 α_{2B}AR) containing a nine-nucleotide deletion, which eliminates three Glu residues from the third intracellular loop of the encoded receptor, displays severely impaired GPCR kinase (GRK)-dependent phosphorylation and is incapable of undergoing desensitization.[20,21] This variant has been genetically associated with acute coronary vasospasm and hypertension in certain populations and under specific conditions, such as stressful work environment.[22–24] Although the effect of this polymorphism on βarrestin binding to the α_{2B}AR has never been directly tested in vitro or in vivo, it is reasonable to surmise that βarrestins fail to interact with, and desensitize, this α_{2B}AR deletion variant, given its complete failure to desensitize. On the other hand, the α_2ARs function as presynaptic inhibitory autoreceptors in the adrenal medulla, similarly to the central nervous system. In other words, they block further catecholamine release from the adrenal glands, a process normally dependent on tonic activation of nicotinic cholinergic receptors by acetylcholine.[25] βarrestin1, acting in concert with GRK2, which is significantly upregulated in the adrenal medulla during HF, causes severe desensitization and downregulation of adrenal α_2ARs, thereby promoting chronic elevation of catecholamine secretion, which underlies the enhanced circulating levels of norepinephrine and epinephrine that accompany and aggravate chronic HF[25–28] (Table 1). Thus, in the case of the adrenal α_2ARs, their desensitization by βarrestin1 plays a detrimental role in HF pathophysiology and blockade of adrenal βarrestin1 action on α_2ARs, either directly or indirectly (by blocking adrenal GRK2), should be pursued for the treatment of chronic HF.

2.2 Cardiac βARs and βarrestins

The principal role of βARs (mainly β_1ARs) in the heart is regulation of cardiac rate and contractility in response to the catecholamines. Out of the three known mammalian βAR subtypes, the β_1AR is by far the predominant one in human adult cardiac myocytes, representing 75–80% of total βAR density, followed by the β_2AR, which comprises about 15–18% of total cardiomyocyte βARs. The remaining 2–3% is β_3ARs.[4,7,19] β_1AR stimulation by catecholamines results in the dissociation of the stimulatory G protein alpha subunit ($G\alpha_s$) from $G_{\beta\gamma}$. $G\alpha_s$ stimulates adenylyl cyclase (AC) to produce cAMP, which, in turn, activates PKA and regulates different intracellular, sarcolemmal and myofibrillar substrates.[4,5,7] Thus, cAMP-dependent

Table 1 In Vivo Cardiovascular Effects of βarrestins, Underlying Molecular Mechanisms, and Therapeutic Implications of Their Targeting for Cardiovascular Disease

Tissue/Receptor	βarrestin Isoform	Signaling Mechanism(s)	In Vivo Cardiovascular Effect(s)	Desirable Intervention (CVD)	References
Adrenal α_2AR	βarrestin1	$G_{i/o}$-protein decoupling	↑ CA secretion	Inhibition (HF)	25–28
Cardiac β_1AR	βarrestin1	G_s-protein decoupling	↓ Cardiac contractility	Inhibition (HF)	29,30
Cardiac β_1- and β_2ARs	βarrestin1	Lack of EGFR transactivation and NFκB inhibition	↑ Cardiac apoptosis/inflammation/post-MI adverse remodeling	Inhibition (HF)	29,30
Cardiac β_1AR (including hArg389 variant)	βarrestin2	SERCA2a SUMOylation/activation; unopposed G_s-protein coupling	↑ Cardiac contractility	Stimulation; carvedilol (?) (HF)	31–34
Cardiac β_1- and β_2ARs	βarrestin2	EGFR transactivation; ERK activation; NFκB inhibition	↓ Cardiac apoptosis/inflammation/post-MI adverse remodeling	Stimulation (HF)	30,31,35,36
Cardiac AT_1R	βarrestin2	EGFR transactivation-ERK-Akt; ↑ Ca^{2+} sensitivity	↑ Cardiac contractility/survival/proliferation	Stimulation (HF, DCM)	37–39
VSM AT_1R	βarrestin2	EGFR transactivation-ERK; BAD inhibition; ↑ DNA synthesis	↑ VSM hyperplasia/hypertrophy	Inhibition (atherosclerosis, arterial stenosis)	40–44
VSM AT_1R	βarrestin1	Unknown; opposed βarrestin2's effects (?)	↓ VSM hyperplasia/hypertrophy	Stimulation (atherosclerosis, arterial stenosis)	40

Adrenal AT_1R	βarrestin1	DGK-PA-ERK-StAR upregulation	↑ Aldosterone synthesis and secretion	Inhibition (HF, hypertension)	45–48
VSM ET_AR	βarrestin2	ERK; p38 MAPK	↑ VSM hyperplasia/hypertrophy/cell migration	Inhibition (atherosclerosis, arterial stenosis)	49
Cardiac fibroblast V_1R	βarrestin1 or βarrestin2	GRK2/ERK	↑ Cardiac fibrosis/adverse remodeling	Inhibition (HF, cardiac post-MI adverse remodeling)	50
Renal collecting duct medullary cell V_2R	βarrestin1 or βarrestin2	G_s-protein decoupling; IGF1R transactivation	↓ Renal H_2O reabsorption	Inhibition (NDI, NSIAD in certain patients with polymorphic V_2R)	51–53
VSM V_1R	βarrestin2	EGFR transactivation-ERK; V_1R:ACKR3 heterodimers	Unknown	Unknown	54,55
VSM GPR109A	βarrestin1	$_cPLA_2$-AA-PGD_2	Niacin-induced vasodilation/"flushing"	Inhibition (niacin's "flushing" adverse effect)	56
Platelet P2YR	βarrestin1	Src-Akt-integrin $\alpha_{IIb}\beta_3$	↑ Thrombosis	Inhibition (TE diseases)	57,58
VSM P2YR	βarrestin1	$G_{q/11}$-protein decoupling	↓ Vasoconstriction/VSM cell migration	Stimulation (hypertension, arterial stenosis)	49,59,60
Platelet PAR4	βarrestin1	PI3K/Akt-fibrinogen binding	↑ Thrombosis	Inhibition (TE diseases)	61,62

Continued

Table 1 In Vivo Cardiovascular Effects of βarrestins, Underlying Molecular Mechanisms, and Therapeutic Implications of Their Targeting for Cardiovascular Disease—cont'd

Tissue/Receptor	βarrestin Isoform	Signaling Mechanism(s)	In Vivo Cardiovascular Effect(s)	Desirable Intervention (CVD)	References
Endothelial PAR4	βarrestin2	GRK5-dependent transactivation by EPCR	↑ Cytoprotection ↓ Inflammation	Stimulation (atherosclerosis, vascular inflammation)	63
Cardiac APJ	βarrestin1 or βarrestin2	Mechanical stretch-dependent (ligand-independent) cytoskeleton mobilization	↑ Cardiac hypertrophy	Inhibition (HF, cardiac maladaptive hypertrophy)	64,65
Endothelial S1P1R	βarrestin2	TNF-dependent NFκB inhibition	↑ Cytoprotection ↓ Inflammation	Stimulation (atherosclerosis, vascular inflammation, HDL cardioprotection)	66

APJ, apelin receptor; *CA*, catecholamine; *CVD*, cardiovascular disease (for which the desirable intervention is intended); *DCM*, dilated cardiomyopathy; *DGK*, diacylglycerol kinase; hArg389 variant, human polymorphic Arg389 β_1AR; *HF*, heart failure; *MAPK*, mitogen-activated protein kinase; *MI*, myocardial infarction; *NDI*, nephrogenic diabetes insipidus; *NSIAD*, nephrogenic syndrome of inappropriate anti-diuresis; *PA*, phosphatidic acid; *StAR*, steroidogenic acute regulatory protein; *TE*, thromboembolic; (?) indicates under investigation or not yet confirmed. See text for all other molecular acronym descriptions.

signaling in cardiomyocytes mediates the cellular effects of β_1AR activation on cardiac chronotropy, inotropy and lusitropy.[4,5,7] As co-factors of GRKs in βAR desensitization/downregulation, βarrestins diminish the inotropic and β-adrenergic reserves of the failing heart and their inhibition should theoretically be beneficial in acute decompensated HF.[4,7] Indeed, genetic deletion of βarrestin1 in the heart results in several desirable therapeutic effects in HF, such as dramatic improvements in both cardiac β-adrenergic and inotropic reserves, amelioration of adverse remodeling and increased survival in βarrestin1-knockout (KO) mice with post-myocardial infarction (MI) HF[29,30] (Table 1). However, contrary to the far more abundant in the heart βarrestin1, βarrestin2 has been shown to be cardioprotective, as it inhibits cardiac apoptosis, inflammation, and significantly attenuates overall adverse remodeling post-MI[35] (Table 1). In addition, β_1AR-activated βarrestin2 was very recently documented to increase cardiac contractility both directly and indirectly[31] (Table 1). The direct effect was mediated through interaction with Sarco/Endoplasmic Reticulum Ca^{2+}-ATPase (SERCA)-2a, leading to enhanced Small Ubiquitin-related MOdifier (SUMO)-ylation of the latter.[31] This process, deficient in human HF, is known to directly stimulate SERCA2a activity, thereby increasing cardiac contractility.[67] βarrestin2 also increases cardiac function indirectly by leaving the β_1AR-stimulated cAMP-dependent pro-contractile signaling intact (i.e., not desensitizing it) in cardiac myocytes in vitro and in post-MI HF mice in vivo.[31] Importantly, these effects in the human heart are not shared by the vastly more abundant βarrestin1.[32]

βarrestin2 also mediates transactivation of the epidermal growth factor receptor (EGFR) by the cardiac β_1AR[30,36] (Table 1). The physiological relevance of this has been demonstrated in transgenic mice overexpressing wild-type β_1ARs or mutant β_1ARs lacking GRK phosphorylation sites and thus, unable to undergo GRK-mediated phosphorylation and bind βarrestins. Under conditions of chronic catecholamine stress, transgenic mice overexpressing the mutant β_1AR are incapable of transactivating cardiac EGFRs and show marked myocyte apoptosis and left ventricular dilatation compared to mice expressing wild-type β_1ARs.[36] Interestingly, several studies have revealed that the beta-blocker carvedilol, an inverse agonist with respect to G protein activation,[19] is a weak βarrestin-biased agonist that activates ERKs (Extracellular signal-Regulated Kinases) via EGFR transactivation[68,69] (Table 1). Unfortunately, these effects of carvedilol have only been demonstrated in heterologous recombinant cell systems with no physiological relevance for the cardiovascular system or for the specific receptor exogenously expressed in them (usually transfected

HEK293 cells). Nevertheless, if these effects occur in cardiomyocytes in vivo, they might represent part of the mechanism for this beta-blocker's cardioprotective effects. Several lines of evidence argue though against this notion. First, sustained β_1AR activation by catecholamines (e.g., isoproterenol or the endogenous natural agonist hormone norepinephrine), which activate βarrestin-dependent ERKs in the heart several times more potently than carvedilol, is cardiotoxic, as it increases cardiac apoptosis, fibrosis, inflammation, adverse remodeling, etc., even in the absence of cardiac injury.[70] In fact, metoprolol, a β_1AR-selective antagonist, was shown to promote cardiac fibrosis and adverse remodeling via βarrestin-stimulated ERK signaling in one study.[71] Moreover, carvedilol binds also β_2ARs and α_1ARs, which may interfere with its βarrestin activating properties,[72] although its cardiac effects are almost exclusively β_1AR-depedent thanks to its relative selectivity for the cardiac β_1AR.[19] Finally, a few recent studies have failed to detect any effects of carvedilol on βarrestin binding to either β_1AR or β_2AR, not to mention that βarrestins appear dispensable for ERK activation by these receptors in some cell systems.[73–75] Even a recent study, which suggested a requirement for $G\alpha_i$ protein recruitment to the carvedilol-bound β_1AR in order for βarrestin binding and EGFR transactivation to follow, could not detect any carvedilol-induced β_1AR–β arrestin interaction.[76] This reported finding is very surprising given that carvedilol is a known G-protein inverse agonist at the cardiac βARs.[19] In contrast, carvedilol was found to selectively recruit βarrestin2 to the hyperfunctional human polymorphic Arg389 β_1AR in transfected NRVMs[33] (Table 1). This human β_1AR variant is significantly more pro-contractile than its Gly389 counterpart, presumably due to more efficient G_s-protein coupling, but, based on the findings of this study, its unique βarrestin2-interacting tropism may play some part in its more robust function (compared to Gly389 β_1AR) as well.[33] Finally, very recent data from our lab indicate that carvedilol-bound β_1AR uniquely stimulates βarrestin2-dependent SERCA2a activity and fractional shortening of NRVMs and of human adult cardiac myocytes in vitro[34] (Table 1). Interestingly enough, the β_2AR seems to counter-regulate this β_1AR/βarrestin2-mediated positive inotropy of carvedilol.[34]

2.3 Other Cardiovascular βARs and βarrestins

It is widely known that β_2ARs are expressed in VSM, wherein they exert vasodilation (VSM relaxation) via intracellular cAMP elevation.[4] VSM cell-residing βarrestins are bound to limit this vasodilation via classic

β_2AR desensitization, resulting in pro-hypertensive states. The fact that GRK2 activity is elevated in human hypertension,[77,78] albeit not known whether it is elevated in VSM per se, argues in favor of vascular βarrestins opposing the VSM-residing β_2AR-dependent vasodilatory effects, since enhanced GRK2 activity generally translates into enhanced βarrestin activity toward GPCRs. However, studies directly testing the effects of βarrestins on β_2AR-mediated vasodilation are lacking and unlikely to be conclusive anyway, given that, remarkably, a vast number of βarrestin substrate GPCRs regulate VSM tone.

Finally, βARs (of both the β_1 and β_2 subtypes) are also present in the adrenal medullae of mice and rats, where, contrary to their α_2AR counterparts, they enhance catecholamine release, acting in an autocrine positive feedback manner (presynaptic facilitatory autoreceptors).[28,79] Therefore, adrenal βarrestin1 should be expected to limit adrenal catecholamine release via desensitizing these receptors. However, α_2ARs play a much more dominant role in regulation of catecholamine secretion in the adrenal medulla, given that they appear to be the only GPCRs that suppress adrenal catecholamine release.[79] α_2ARs can also upregulate GRK2 transcriptionally in adrenal chromaffin cells, which adds another layer of regulation of α_2AR-gated catecholamine secretion.[80] Thus, adrenal βarrestin1 actions on α_2ARs (rather than on βARs) are the determining factor in regulation of catecholamine secretion from the adrenal medulla.[81]

3. CARDIOVASCULAR ANGIOTENSIN II RECEPTORS AND βARRESTINS

3.1 Cardiac AT_1Rs and βarrestins

The AT_1R plays important regulatory roles in the heart, mainly via actions in cardiac fibroblasts and endothelial cells, but, to a much lesser extent, also in cardiac myocytes.[4,7,82–84] However, its abundance in the adult human myocardium (either failing or non-failing) is remarkably low, especially compared to the predominant cardiac GPCR type β_1AR (density ratio of AT_1R/β_1AR in the non-failing human heart: ~1:15).[19] The discovery of the AngII peptide analog SII ([Sar^1-Ile^4-Ile^8]-AngII), which, when bound to the AT_1R, elicits βarrestin signaling but no G_q protein signaling,[9] pioneered the research on cardiac biased AT_1R signaling. However, we now know that SII is not completely βarrestin-"biased," as it can actually elicit signaling via other G-protein types (e.g., G_s, G_i) from the AT_1R, i.e., in reality, is a G_q protein-"biased" inverse agonist.[85] Nevertheless, studies have shown that AT_1R-elicited βarrestin-dependent signaling in cardiac myocytes leads to cardiomyocyte proliferation without hypertrophy, which is $G_{q/11}$ protein signaling-dependent, and can even result in positive inotropy and lusitropy depending on which GRK isoform has phosphorylated the receptor (the so-called receptor "bar-coding" by GRKs) but also on which βarrestin isoform is engaged.[37,86] Specifically, GRK2 and βarrestin1 seem to reduce, whereas GRK6 and βarrestin2 seem to promote AT_1R-induced contractility in primary murine adult cardiomyocytes.[37] In isolated Langendorff-perfused cardiac preparations however, AT_1R-activated βarrestins do not exert inotropic or chronotropic effects, despite the fact they still induce ERK signaling.[87] Additionally, a recombinant $AT_{1A}R$ capable of activating only βarrestins inhibits myocardial apoptosis and fibrosis, and enhances cardiomyocyte hypertrophy, bradycardia, and fetal cardiac gene expression upon transgenic overexpression in murine cardiomyocytes.[38] Finally, a AT_1R βarrestin2-"biased" ligand was recently reported to exert beneficial effects in a murine model of dilated cardiomyopathy via both prevention of maladaptive signaling and improved myofilament Ca^{2+} sensitivity[39] (Table 1). Thus, it seems that, while cardiac AT_1R promotes hypertrophy and cardiomyocyte proliferation via the classic G_q protein/phospholipase C signaling pathway, it can increase cardiac function via βarrestin-dependent signaling, specifically βarrestin2-dependent (Table 1).

Several SII-derivative peptides, i.e., βarrestin-"biased" AT_1R agonists, developed by the pharmaceutical company Trevena Inc., showed initially promising results in HF animal models via a combination of $G_{q/11}$ protein-mediated vasoconstriction blockade and promotion of βarrestin-dependent signaling in cardiac myocytes.[88–90] For instance, TRV120027 blocked the G_q protein-dependent vasoconstriction induced by the AT_1R, thereby reducing cardiac afterload to boost cardiac function, and also inhibited AT_1R-dependent cardiac hypertrophy that leads to HF.[90] A similar βarrestin-"biased" AT_1R agonist peptide, TRV120026, shared these effects, as well.[9,88] However, in a phase III clinical trial for acute decompensated HF, these compounds failed to demonstrate clinical benefits and their clinical development was terminated (BLAST-AHF, ClinicalTrials.gov number, NCT01966601). The reasons for this failure are unclear; it might have been simply the fact that findings in animal models are not always translatable to humans or that the compounds were not completely βarrestin-"biased." Another explanation might be that the cardioprotective βarrestin2 is markedly under-expressed in the human heart,[32] so these compounds essentially stimulated the cardiotoxic βarrestin1 in the patients' hearts. Alternatively, these biased ligands may have interacted with the AT_1R in cardiac fibroblasts instead of myocytes, and thus, no clinical benefit could be attained.

Finally, cardiac AT_1R-induced βarrestin-dependent signaling has been shown to underlie mechanical stretch-activated AT_1R.[91] More specifically, unliganded (agonist-unbound) $AT_{1A}R$ has been reported to selectively activate βarrestins, instead of G-proteins, in response to simple mechanical stretch.[91] The stretch-activated $AT_{1A}R$-elicited βarrestin signaling results in EGFR transactivation via ERKs and Akt, promoting cardiomyocyte anti-apoptosis.[91] In the same vein, a recent study suggested that AT_1R-induced βarrestin-dependent signaling also underlies the Frank–Starling mechanism of cardiac contractility.[92] This finding is quite intriguing, given that (a) this mechanism is not known to be driven by a particular hormone/receptor but rather is hemodynamically governed (i.e., it is based on the dynamic balance between cardiac preload and afterload and on myocardial wall elasticity), and (b) even if it is dependent on a particular hormone/receptor, then the most likely candidate receptor for that would be the β_1AR, which is by far the most abundant GPCR in (at least human) cardiomyocytes, directly increasing myofilament Ca^{2+} currents through cAMP/PKA.[19]

3.2 Vascular AT_1Rs and βarrestins

AngII acting through the AT_1R is not only one of the most potent vasoconstrictive but also one of the most powerful growth signals for VSM cells, stimulating cell proliferation and causing VSM hyperplasia, a major contributing factor in vascular restenosis (for instance, after percutaneous coronary intervention followed by stent placement) and in atherosclerosis.[93] VSM cell-residing AT_1R induces βarrestin-dependent signaling that contributes to pro-atherosclerotic effects.[40] Specifically, neointimal hyperplasia after carotid endothelial injury is enhanced in βarrestin1KO mice but, notably, diminished in βarrestin2KO mice.[40] Loss of βarrestin2 appears to decrease GPCR-stimulated VSM cell proliferation and ERK1/2-dependent activation/migration, consistent with a role for βarrestin2-dependent signaling in the injury response.[40] In vitro, both G protein- and βarrestin-dependent pathways elicited by the $AT_{1A}R$ converge on EGFR transactivation in primary VSM cells to stimulate proliferation.[41] Specifically, βarrestin2 seems to enhance non-receptor Src protein kinase-dependent EGFR transactivation by the $AT_{1A}R$.[41] Moreover, βarrestin2-dependent ERK activation downregulates the pro-apoptotic phospho-BAD (Bcl-2-associated death promoter) protein, thus inducing anti-apoptotic and cytoprotective effects in rat VSM cells.[94] Other signaling mechanisms mediating the stimulatory effects of AT_1R-activated βarrestin2 on VSM cell proliferation and migration, thereby contributing to vascular restenosis/atherosclerosis and neointimal formation, include activation of the small GTPase (monomeric G-protein) ADP-ribosylation factor (ARF)-6[42] and nuclear factor (NF)-kappaB-dependent cyclooxygenase (COX)-2 upregulation.[43,44,95,96] Thus, it appears that βarrestin2 can promote AT_1R-dependent VSM proliferation and hypertrophy, thereby contributing to the development and progression of atheromas and vascular (re)stenosis, whereas βarrestin1 may oppose these deleterious (in VSM) actions of βarrestin2 (Table 1). This is in complete agreement with a very recent study reporting that βarrestin2 promotes neointimal formation and VSM cell migration/restenosis in injured arteries by mediating 5-HT_{2B} serotonergic receptor-induced mammalian target of rapamycin (mTOR)/p70/p85 ribosomal S6-kinase (S6K) signaling,[97] and with an older study reporting that βarrestin-dependent signaling from the AT_1R results in stimulation of protein synthesis via Akt kinase-mTOR-p70/85S6K and ERK-p90 ribosomal S6-kinase (RSK) pathways in primary VSM cells.[95] Interestingly, the AT_1R reportedly heterodimerizes with purinergic P2Y6 receptors in VSM cells.[98] This heterodimerization switches AT_1R signaling from βarrestin-dependent pathways promoting proliferation to G-protein-dependent pathways promoting hypertrophy and contributes to AngII-induced hypertension in mice.[98] On the other hand, βarrestin2, acting upon AT_1R:bradykinin B2 receptor heterodimers, can inhibit B2 receptor signaling in primary VSM cells.[96] Additionally, the net effect of βarrestin2 on vascular inflammation may depend on its ubiquitination status in VSM cells, which is regulated by ubiquitin-specific protease (USP)-20.[99]

To sum up, the two βarrestins appear to play swapped roles in the VSM vs. the myocardium: in VSM, βarrestin2 appears to exert toxic, pro-atherotic actions via promotion of VSM cell proliferation and βarrestin1 appears beneficial by opposing these effects. In contrast in the heart, βarrestin1 is the damaging isoform by exerting negative inotropy and promoting apoptosis, inflammation, and overall adverse remodeling post-MI, whereas βarrestin2 is the beneficial βarrestin, promoting contractility post-MI and in dilated cardiomyopathy, as well as anti-apoptosis, anti-inflammation, and overall reverse remodeling post-MI (Table 1). The reason for this might lie in the different GPCRs that induce these effects of βarrestins in these two different cardiovascular tissue types: in the heart, βarrestins regulate contractility via βARs (mainly β_1ARs), whereas in VSM, they regulate cell proliferation, migration, and inflammation (mainly) via the AT_1R (Table 1).

3.3 Adrenal AT_1Rs and βarrestins

The adrenal AT_1R stimulates synthesis and secretion of the cardiotoxic hormone aldosterone, elevated in chronic human HF, via both $G_{q/11}$ protein- and βarrestin1-dependent signaling[82,100] (Table 1). In adrenocortical zona glomerulosa cells, βarrestin1-dependent signaling from the AT_1R promotes aldosterone synthesis and secretion both in vitro and in vivo and inhibition of this pathway in the adrenal cortex in vivo markedly attenuates post-MI HF progression.[45,46] In fact, βarrestin1 is indispensable for adrenal aldosterone production, since βarrestin1KO mice have normal circulating aldosterone levels even in the presence of post-MI HF.[30] Since AngII-dependent aldosterone production proceeds through two independent signal transducers, G-proteins and βarrestins, it follows that the activation of both needs to be prevented for complete aldosterone suppression post-MI. To that end, the currently FDA-approved AT_1R antagonist drugs (sartans or angiotensin receptor blockers, ARBs) were recently screened for βarrestin-dependent aldosterone suppression. The prototypic agent of the class, losartan, was found to be a relatively G protein-"biased" antagonist, leaving βarrestin1-dependent aldosterone production largely unaffected.[46–48] Interestingly however, its active metabolite EXP1374 is effective at blocking AT_1R-dependent aldosterone production and βarrestin activation.[48] Regarding the rest of the sartans, the G-protein inverse agonists candesartan and valsartan display significant inverse agonism toward βarrestins as well, so they are, in effect, unbiased inverse agonists for both G-proteins and βarrestins.[47,48] This makes them the most efficacious aldosterone suppressors both in vitro and in vivo.[47,48] In contrast, irbesartan is a G protein-"biased" inverse agonist, displaying very weak βarrestin inhibition and aldosterone suppression.[47,48] Based on these findings, we designed a compound with a carboxyl-benzodiazole moiety and a long aliphatic chain substituent, connected via the methylene carbon to the tetrazolo-biphenyl-methylene backbone of sartans.[48] Preliminary in vitro data show that, as expected, it is a potent βarrestin-"biased" inverse agonist at the AT_1R conferring significant aldosterone suppression (Ref. 100; A. Lymperopoulos, unpublished data). Of course, ARBs are orthosteric AT_1R antagonists, so the βarrestin inhibition they produce is theoretically equivalent to the extent of their antagonism of AngII-AT_1R

binding. Thus, testing them for βarrestin antagonism/inverse agonism migh seem pointless. However, even the unliganded AT_1R can activate βarrestin signaling toward ERK activation, albeit not through the classic GRK-dependent, but rather via protein kinase C (heterologous desensitization/phosphorylation).[101] Besides, mechanical stretch-activated, i.e., unliganded, AT_1R does the same thing (see Ref. 40; Section 3.1). Therefore, it is important to know the inverse agonist properties of the ARBs regarding not only G-protein but also βarrestin activation at the AT_1R.

Unlike the receptor's structural requirements for βarrestin activation, for which information is still very limited, several structural features of the AngII peptide hormone itself necessary for AT_1R "bias" toward βarrestins have been elucidated. Specifically, the aromatic side-chain of Phe8, although indispensable for full G protein agonism, is not required for βarrestin agonism.[102] Tyr4 is also essential for full G_q-protein agonism but does not affect receptor phosphorylation, which is usually required for βarrestin activation.[103] Interestingly, the size of the aliphatic side-chain of Ile5 was recently found to be directly proportional to the amount of βarrestin agonism.[104] Therefore, the stereochemical "bulkiness" of the Ile5 position might play an important auxiliary role in βarrestin activation by the AT_1R.

Finally, AT_1R-activated βarrestin1 (but, interestingly, not βarrestin2) was recently shown to promote acute stimulation of catecholamine secretion in adrenomedullary chromaffin cells, via direct interaction with plasma membrane-residing transient receptor potential channel (TRPC) type 3 (TRPC3) Ca^{2+} channels.[105] Thus, it is quite plausible that AT_1R-bound βarrestin1 can directly stimulate TRPC3-dependent Ca^{2+} currents in the adrenal cortex, as well, thereby inducing AngII-dependent aldosterone secretion within seconds of agonist binding (and in parallel to AT_1R-elicited G_q-protein signaling). This may represent another signaling mechanism by which βarrestin1 can promote adrenocortical aldosterone production.

4. OTHER CARDIOVASCULAR GPCRs AND βARRESTINS

4.1 Endothelin Receptors

Overstimulation of endothelin (ET)-1 type A (ET_A) receptors, which couple primarily with G_q-proteins in VSM cells, causes vasoconstriction, vascular hypertrophy and occlusion.[106] Additionally, VSM cell-residing ET_A receptors play important roles in cell proliferation and migration, both of which underlie neointimal formation in atherosclerotic plaques promoting atherogenesis.[106] In rat aortic smooth muscle cells, βarrestin2 desensitizes

ET_A receptors, while it promotes VSM cell migration via ET_A receptor-dependent p38 MAPK and ERK signals.[49] Thus, βarrestin2 is an integral component of VSM remodeling and stenosis by transducing intracellular signals not only from AngII but also from ET-1 in VSM cells (Table 1).

4.2 Vasopressin Receptors

Arginine vasopressin (AVP) is another important hormone for the cardiovascular system that exerts effects through two different GPCR types, V_1 and V_2.[107] V_1 vasopressin receptor (V_1R) is a $G_{q/11}$-coupled receptor activating phospholipase C and located in VSM cells, wherein it promotes vasoconstriction, whereas V_2R is a G_s protein-coupled receptor in epithelial cells, mainly the principal cells of the distal nephron, where it stimulates renal water reabsorption causing hyponatremia via intracellular cAMP elevation.[107] Both vasopressin receptor types interact with βarrestins in vitro and, in fact, V_2R is the prototypic example of a class B GPCR in terms of βarrestin interaction kinetics, i.e., binds to, and internalizes with both βarrestins equally well, and the βarrestin interaction is strong and sustained.[51,108] In fact, a polymorphic human V_2R variant (Arg137His), strongly associated with nephrogenic diabetes insipidus (NDI) in afflicted carriers, undergoes constitutive βarrestin-mediated desensitization and internalization resulting in a loss-of-function receptor[52] (Table 1). The cardiac fibroblast-residing V_1R was recently shown to promote cardiac fibroblast proliferation and, consequently, cardiac fibrosis, via ERK activation mediated by GRK2 and either βarrestin isoform[50] (Table 1). Similarly, ERK activation by the V_2R in rat kidney medullar collecting ducts in vivo requires βarrestins but, intriguingly, these βarrestins seem to get activated by the insulin-like growth factor-1 receptor (IGF1R), a receptor tyrosine kinase transactivated by the V_2R in renal medullary cells[53] (Table 1). Finally, regarding VSM cells, the V_1R has also been reported to activate ERKs via βarrestin2 and EGFR transactivation in a rat VSM cell line,[54] while in human VSM cells, it forms heterodimers with the atypical chemokine receptor (ACKR)-3, and this heterodimerization reduces AVP-induced βarrestin2 recruitment to the V_1R[55] (Table 1).

4.3 Niacin Receptor (GPR109A)

Niacin (nicotinic acid), recently fallen out of favor due to lack of clinical efficacy,[109] has been used as a lipid-lowering drug for over 50 years with one big caveat: it causes severe "flushing," which can limit its dosing and tolerability in patients.[109] The $G_{i/o}$ protein-coupled receptor GPR109A is responsible for both the niacin's therapeutic effect and adverse effect of "flushing."[110] Whereas the lipid-lowering effect of GPR109A is G-protein-mediated, the "flushing" is reportedly due to vascular βarrestin1-mediated activation of cytosolic phospholipase ($_c$PL)-A_2, which causes release of arachidonic acid (AA), the precursor of all prostaglandins, including prostaglandin D_2 (PGD_2). The latter dilates the cutaneous vessels via its DP_1 G protein-coupled receptor causing the unpleasant "flushing" effect.[56] Thus, a G-protein-"biased" agonist at the GPR109A would be therapeutically desirable, as it would abolish the need for combining niacin with a DP_1 receptor antagonist (e.g., laropiprant), which is what current clinical practice guidelines mandate (Table 1). In that vein, 4-(phenyl) thio-1*H*-pyrazole has been validated as a pharmacophore scaffold for GPR109A agonist activity and 1-(pyrazin-2-oyl)-derivatives have been synthesized with promising G-protein-"biased" agonist properties at the GPR109A receptor.[111]

4.4 P2Y Receptors

The P2Y family of purinergic receptors (receptors for purine nucleosides and nucleotides, such as adenosine 5′-diphosphate, ADP, and adenosine 5′-triphosphate, ATP) constitutes the metabotropic purinergic receptor family, whose members are all GPCRs.[112] $P2Y_1$ and $P2Y_{12}$ receptors for ADP are expressed in platelets and play key roles in platelet activation and thrombosis.[112] Platelets activated by ADP become refractory to re-stimulation, a phenomenon which has been attributed to $P2Y_1$ receptor desensitization and internalization.[113] Although a substantial fraction of $P2Y_{12}$ receptors is also rapidly internalized upon stimulation, whether this ADP receptor type is also subject to desensitization is a matter of debate.[113] Platelet βarrestin1 is known to promote thrombus formation by mediating signaling of P2Y receptors to integrin $\alpha_{IIb}\beta_3$, while leaving P2Y receptor desensitization unaffected.[57] This βarrestin1-dependent signaling to integrin $\alpha_{IIb}\beta_3$ is Src- and Akt kinase-dependent.[57] Of note, it was very recently reported that the $P2Y_{12}$ receptor heterodimerizes with protease-activated receptor (PAR)-4 (see Section 4.5) in platelets, and this heterodimerization regulates βarrestin1-dependent Akt signaling.[58] Thus, P2Y receptor-activated βarrestin1 in platelets can promote arterial thrombosis (Table 1), potentially in a PAR4-regulated manner, so platelet βarrestin1 might constitute a plausible therapeutic target in thromboembolic disease.

P2Y receptors are also present in VSM cells, wherein they induce vasoconstriction, hypertrophy and hyperplasia, via classic coupling to the $G_{q/11}$-protein-phospholipase C-Ca^{2+} signaling pathway.[59] βarrestin1, but not βarrestin2, has been shown to classically desensitize P2Y receptor-induced pro-contractile signaling in adult Wistar rat mesenteric arterial smooth muscle cells in vitro, in response to uridine 5′-triphosphate (UTP), a purine nucleotide.[49,59,60] In addition, βarrestin1 desensitizes UTP-stimulated P2Y receptor signaling to p38 MAPK and ERK activation in rat aortic smooth muscle cells.[49] Given that the AT_1R:P2Y6 receptor heterodimers found to form in VSM cells are βarrestin recruitment-deficient and contribute to AngII-induced hypertension in mice (Ref. 98; see also Section 3.2), it seems quite plausible that VSM cell βarrestin1 promotes vasodilatation, at least upon acting on P2Y receptors (Table 1). Therefore, contrary to platelets, wherein P2Y receptor-activated βarrestin1 promotes thrombosis, βarrestin1 may exert beneficial (in hypertension, heart failure, etc.) vasodilation in VSM cells by opposing (desensitizing) P2Y receptor-induced vasoconstriction (Table 1).

4.5 Protease-Activated Receptors

Another family of GPCRs with important effects for the cardiovascular system is the protease-activated receptor (PAR) family, which responds to thrombin and other proteases courtesy of proteolytic cleavage of the receptor's extracellular N-terminus.[114] They are divided into four distinct types (PAR1–4), all coupling primarily to the $G_{q/11}$-protein-phospholipase C-Ca^{2+} signaling pathway (PAR1 and PAR2 can also couple to G_i-proteins, i.e., to inhibition of cAMP). PARs reside in plasma membranes of vascular endothelial cells and platelets, among other cell types, wherein they mediate the well-characterized process of blood clot formation.[114] PARs are βarrestin substrates with βarrestin1 mediating the signaling from thrombin-stimulated platelet PAR4, in addition to its role in ADP/P2Y receptor signaling in this cell type (see also Section 4.4). More specifically, βarrestin1 (but not βarrestin2) mediates PAR4 signaling to phosphatidylinositol-3 kinase (PI3K) and subsequent Akt kinase activation in platelets, which increases fibrinogen binding and thrombus formation[61,62] (Table 1). ADP can positively modulate this βarrestin1 effect in platelets via $P2Y_1$ receptors that activate PKC and $P2Y_{12}$ receptors that dimerize with PAR4 (see also Section 4.4).[58,61,62] Thus, βarrestin1 (but not βarrestin2) promotes thrombosis from both purinergic and thrombin receptors (PARs) in platelets (Table 1).

In contrast, some very interesting findings have been reported for the vascular endothelial cell-residing PAR1 and its βarrestin signaling properties. Activated protein C (APC) and thrombin elicit paradoxical cytoprotective and cytotoxic signaling responses in vascular endothelial cells through cleavage of PAR1 at Arg46 and Arg41 protease recognition sites, respectively.[115] On the other hand, PAR1 signaling via βarrestin2 has been shown to be cytoprotective and anti-inflammatory in vascular endothelial cells.[116] It was recently shown that both thrombin and APC can elicit cytoprotective/anti-inflammatory PAR1 signaling in vascular endothelial cells via βarrestin2 (βarrestin2-"biased" agonism).[115] Notably, this PAR1-dependent βarrestin2-"biased" signaling did not result from protease-mediated PAR1 protein cleavage but instead, APC and thrombin bound the endothelial protein C (APC) receptor (EPCR) to recruit GRK5, followed by transactivation of PAR1-elicited βarrestin2-dependent signaling[63] (Table 1). These studies indicate that vascular endothelial cell-residing PAR-activated βarrestin2 is cytoprotective and anti-inflammatory, i.e., beneficial for the systemic vasculature, in complete contrast with platelet PAR (or P2Y receptor)-activated βarrestin1, which is pro-thrombotic in the systemic vasculature (Table 1).

4.6 Apelin Receptor

The apelin receptor (APJ), an AT_1R-like GPCR that does not respond to AngII,[64,117] has been shown to induce vasodilation, increasing cardiac output via afterload reduction, through G-protein signaling in vascular endothelial cells.[118–120] In contrast, mechanical stretch-activated cardiac apelin receptor causes cardiac hypertrophy via βarrestins (Table 1), which is inhibited by G-protein signaling.[64,65] For this reason, development of cardiac apelin G-protein-"biased" agonists might be of therapeutic value and is currently actively pursued.[121]

4.7 Sphingosine-1-Phosphate 1 Receptor

The chronic inflammation known to underlie atherosclerosis and other cardiovascular diseases is modulated by sphingosine-1-phosphate (S1P). This bioactive phospholipid activates its S1P1 receptor (S1P1R), a GPCR that also signals through βarrestins.[122] Although S1P1R normally promotes inflammation via G-protein signaling,[123–125] it can actually inhibit systemic chronic inflammation in vascular endothelial cells if activated by high-density lipoprotein (HDL)-bound S1P thanks to βarrestin2 recruitment.[66] The precise mechanism involves suppression of tumor necrosis factor (TNF)-dependent NFκB activation by βarrestin2,[66] which thus acts as anti-inflammatory in exactly the same manner as it does in the myocardium (Refs. 31, 35; see also Section 2.2) (Table 1). Therefore, vascular endothelial S1P1R-activated βarrestin2 attenuates atherosclerosis-associated vascular inflammation, which may be partly responsible for the cardioprotective actions of HDL (the "good" cholesterol) (Table 1). Consequently, whether activated by PARs or by S1P1R, βarrestin2 is cytoprotective and anti-inflammatory in vascular endothelial cells (as well as in the myocardium).

5. THERAPEUTIC IMPLICATIONS OF THE FUNCTIONAL DIVERGENCE OF CARDIOVASCULAR βARRESTINS

It becomes clear from the studies reviewed here that there is a remarkable functional divergence between the two βarrestins in almost every cardiovascular tissue/cell type studied, to the point that in some parts of the circulatory system, e.g., myocardium or VSM, the two isoforms directly oppose each other's actions. This is true for several other mammalian organ systems and tissue types as well.[126,127] Therefore, the two βarrestins that were once thought of as functionally interchangeable, since they both desensitize and internalize GPCRs, are now clearly recognized as functionally different, at least in terms of their signal transducing properties.[126,127] This inter-βarrestin isoform functional variation is supported by molecular, biophysical, crystallographic, and proteomic lines of evidence.[128–131] From the perceptive of evolutionary biology, it also makes perfect sense: if βarrestin1 and βarrestin2 performed identical functions in every cell and tissue, one of them would be biologically redundant and would have long been lost in the evolutionary history of mammals. Of course, the fact that only the double βarrestin1/2KO mouse is embryonic-lethal argues for some biological redundancy between the two isoforms. Perhaps the two βarrestins can compensate for each other during embryonic development but not later in adult life.

In any case, the two βarrestins display significant functional diversity in the cardiovascular system, which can have enormous therapeutic implications (summarized in Table 1). First in the heart, βarrestin1 exerts negative inotropy, by desensitizing the cAMP-dependent pro-contractile signaling of the β_1AR, and promotes cardiac apoptosis, inflammation, and other adverse remodeling-associated processes post-MI. In direct contrast to these overall negative effects of βarrestin1, particularly in the context of HF, βarrestin2 promotes β_1AR-mediated cardiac contractility directly, via SERCA2a potentiation, and indirectly, by leaving β_1AR's cAMP-dependent pro-contractile signaling unaffected. At the same time, it inhibits apoptosis and inflammation, and attenuates adverse remodeling of the failing heart via EGFR transactivation, NFκB inhibition, ERK activation, and other mitogenic/proliferative molecular signaling pathways in cardiomyocytes. Thus, myocardial βarrestin2 stimulation or βarrestin1 inhibition may have therapeutic potential in HF.

The exactly inverse situation occurs in the VSM. Here, the actions of βarrestin1 appear beneficial in terms of vasodilation and attenuation of hyperplasia, courtesy of opposing the $G_{q/11}$-protein-dependent pro-hypertrophic signaling of AT_1Rs and P2YRs. Conversely, βarrestin2 seems to promote VSM hypertrophy/hyperplasia via pro-hypertrophic and anti-apoptotic signaling to EGFR transactivation, MAPK activation, and BAD inhibition when activated by AT_1Rs and ET_ARs. Thus, in the VSM, βarrestin2 inhibition or βarrestin1 stimulation may be pursued for arterial stenosis or atherosclerosis treatment. The sharp functional contrast between cardiac and VSM βarrestin functions is probably due to the different GPCRs activating them in each of these types of cardiovascular tissue. In the heart, βarrestin2 is beneficial and βarrestin1 toxic due to their differential effects on β_1AR (and, to a lesser extent, β_2AR) signaling but in the VSM, they have exchanged roles (i.e., βarrestin1 is beneficial and βarrestin2 is toxic) due to their differential effects on (primarily) AT_1R signaling. Pharmacologically, this is not surprising at all; the exact same phenomenon occurs with the major second messenger cAMP in smooth muscle cells vs. cardiomyocytes, as well: cAMP is pro-contractile in the myocardium in response to β_1ARs but pro-relaxing in the VSM in response to β_2ARs.[4]

Finally, in vascular endothelial cells and in platelets, βarrestin2 appears beneficial by promoting cellular protection and suppressing inflammation in the former cell type, while βarrestin1 is detrimental for the systemic circulation by promoting thrombosis in the latter. Since nothing is known about the respective counterpart isoforms in these two cell types, i.e., βarrestin1 in endothelial cells and βarrestin2 in platelets, endothelial βarrestin2 stimulation and platelet βarrestin1 inhibition appear plausible pharmacological approaches for the treatment of atherosclerosis and thromboembolic disease, respectively.

6. CONCLUSIONS AND FUTURE PERSPECTIVES

Considerable progress has been made over the past two decades in delineating βarrestin-dependent signaling pathways and the resultant effects in all cardiovascular tissues and cell types in vitro and in vivo. Unfortunately, however, the vast majority of the studies done so far are in vitro in cellular systems usually expressing recombinant (non-native) GPCRs at supraphysiological levels. Very few data are available from in vivo studies in heart disease animal models, let alone patients. This may considerably hamper the translational impact of the field of cardiovascular βarrestin pharmacology. Nevertheless, it has become abundantly clear that the following three factors have absolutely to be considered, regarding the therapeutic potential of cardiovascular βarrestin targeting: (a) which tissue/cell type (e.g., cardiac myocyte, cardiac fibroblast, VSM cell, vascular endothelial cell, platelet, etc.) is targeted; (b) which βarrestin isoform (βarrestin1 or βarrestin2) is targeted; and (c) which GPCR is targeted.

Although more studies, especially in vivo, are warranted to fully bring the picture of cardiovascular βarrestin physiology into clear focus, some conclusions can be drawn already from the data available to date. For instance, in the heart, βarrestin1 is detrimental but βarrestin2 is beneficial. Given the latter is significantly under-expressed in the (non-failing) human myocardium, cardiac-specific βarrestin2 gene transfer (or pharmacological stimulation) for heart failure therapy seems an elegant approach. Similarly, βarrestin2 appears overall beneficial also in the vascular endothelium, where it suppresses inflammation and mediates cytoprotection, including part of HDL's cardioprotective effects, although some of these benefits may get obscured by βarrestin2's apparent actions on AT_1Rs and ET_ARs in VSM cells. In contrast, βarrestin1 seems to exert overall negative effects in the cardiovascular system: it promotes thrombosis in platelets, opposes the beneficial water/electrolyte balancing properties of arginine vasopressin in the kidneys, promotes AngII-dependent aldosterone production/secretion by the adrenal gland, and is even responsible for niacin's main adverse effect of "flushing." Only in VSM cells does βarrestin1 seem to result in beneficial actions, as it purportedly attenuates vasoconstriction and vascular hyperplasia/hypertrophy, most likely thanks to its classic desensitization of AT_1R's and P2YR's $G_{q/11}$-protein-mediated signaling. Thus, the evidence accumulated so far seems to favor βarrestin1-blocking and/or βarrestin2-enhancing approaches for the treatment of cardiovascular disease, whether it be heart failure, hypertension, atherosclerosis, thromboembolism, or even cardio-renal disorders (e.g., nephrogenic diabetes insipidus). The hope is that future in vivo studies will clarify in full the exact roles of the individual βarrestin isoforms in each component of the cardiovascular system and that at least some of the findings will be validated in humans, as well. Only when that happens, will we be able to gauge with certainty the true potential of cardiovascular βarrestins as therapeutic targets. This will also decide whether the current "hype" of GPCR "biased" agonism for cardiovascular drug discovery will eventually become reality or get buried into oblivion definitively.

REFERENCES

1. Desimine VL, McCrink KA, Parker BM, Wertz SL, Maning J, Lymperopoulos A. Biased agonism/antagonism of cardiovascular GPCRs for heart failure therapy. *Int Rev Cell Mol Biol.* 2018;339:41–61.
2. Hilger D, Masureel M, Kobilka BK. Structure and dynamics of GPCR signaling complexes. *Nat Struct Mol Biol.* 2018;25:4–12.
3. Peterson YK, Luttrell LM. The diverse roles of arrestin scaffolds in G protein-coupled receptor signaling. *Pharmacol Rev.* 2017;69:229–256.
4. Lymperopoulos A, Rengo G, Koch WJ. Adrenergic nervous system in heart failure: Pathophysiology and therapy. *Circ Res.* 2013;113:739–753.
5. Lymperopoulos A, Garcia D, Walklett K. Pharmacogenetics of cardiac inotropy. *Pharmacogenomics.* 2014;15:1807–1821.
6. Siryk-Bathgate A, Dabul S, Lymperopoulos A. Current and future G protein-coupled receptor signaling targets for heart failure therapy. *Drug Des Devel Ther.* 2013;7:1209–1222.
7. Capote LA, Mendez Perez R, Lymperopoulos A. GPCR signaling and cardiac function. *Eur J Pharmacol.* 2015;763:143–148.
8. Lymperopoulos A, Bathgate A. Arrestins in the cardiovascular system. *Prog Mol Biol Transl Sci.* 2013;118:297–334.
9. Whalen EJ, Rajagopal S, Lefkowitz RJ. Therapeutic potential of β-arrestin- and G protein-biased agonists. *Trends Mol Med.* 2011;17:126–139.
10. Ranjan R, Dwivedi H, Baidya M, Kumar M, Shukla AK. Novel structural insights into GPCR-β-arrestin interaction and signaling. *Trends Cell Biol.* 2017;27:851–862.
11. Eichel K, Jullié D, Barsi-Rhyne B, et al. Catalytic activation of β-arrestin by GPCRs. *Nature.* 2018;557:381–386.
12. Piascik MT, Perez DM. Alpha1-adrenergic receptors: new insights and directions. *J Pharmacol Exp Ther.* 2001;298:403–410.
13. Jensen BC, Swigart PM, De Marco T, Hoopes C, Simpson PC. {alpha}1-adrenergic receptor subtypes in nonfailing and failing human myocardium. *Circ Heart Fail.* 2009;2:654–663.
14. Van Tassell BW, Rondina MT, Huggins F, Gilbert EM, Munger MA. Carvedilol increases blood pressure response to phenylephrine infusion in heart failure subjects with systolic dysfunction: Evidence of improved vascular alpha1-adrenoreceptor signal transduction. *Am Heart J.* 2008;156:315–321.
15. Stewart JM, Patel MB, Wang J, et al. Chronic elevation of norepinephrine in conscious dogs produces hypertrophy with no loss of LV reserve. *Am J Physiol.* 2009;262: H331–H339.

16. Vecchione C, Fratta L, Rizzoni D, et al. Cardiovascular influences of alpha1b-adrenergic receptor defect in mice. *Circulation.* 2002;105:1700–1707.
17. Albee LJ, Eby JM, Tripathi A, et al. α_1-Adrenergic receptors function within hetero-oligomeric complexes with atypical chemokine receptor 3 and chemokine (C-X-C motif) receptor 4 in vascular smooth muscle cells. *J Am Heart Assoc.* 2017;6(8).
18. Shi Q, Li M, Mika D, et al. Heterologous desensitization of cardiac β-adrenergic signal via hormone-induced βAR/arrestin/PDE4 complexes. *Cardiovasc Res.* 2017; 113:656–670.
19. Yoshikawa T, Port JD, Asano K, et al. Cardiac adrenergic receptor effects of carvedilol. *Eur Heart J.* 1996;17(Suppl B):8–16.
20. Small KM, Brown KM, Forbes SL, Liggett SB. Polymorphic deletion of three intracellular acidic residues of the alpha2B-adrenergic receptor decreases G protein-coupled receptor kinase-mediated phosphorylation and desensitization. *J Biol Chem.* 2001;276:4917–4922.
21. Nguyen K, Kassimatis T, Lymperopoulos A. Impaired desensitization of a human polymorphic α2B-adrenergic receptor variant enhances its sympatho-inhibitory activity in chromaffin cells. *Cell Commun Signal.* 2011;9:5.
22. Snapir A, Mikkelsson J, Perola M, Penttilä A, Scheinin M, Karhunen PJ. Variation in the alpha2B-adrenoceptor gene as a risk factor for prehospital fatal myocardial infarction and sudden cardiac death. *J Am Coll Cardiol.* 2003;41:190–194.
23. Von Wowern F, Bengtsson K, Lindblad U, Råstam L, Melander O. Functional variant in the (alpha)2B adrenoceptor gene, a positional candidate on chromosome 2, associates with hypertension. *Hypertension.* 2004;43:592–597.
24. Ohlin B, Berglund G, Nilsson PM, Melander O. Job strain, decision latitude and alpha2B-adrenergic receptor polymorphism significantly interact, and associate with high blood pressures in men. *J Hypertens.* 2007;25:1613–1619.
25. Lymperopoulos A, Rengo G, Funakoshi H, Eckhart AD, Koch WJ. Adrenal GRK2 upregulation mediates sympathetic overdrive in heart failure. *Nat Med.* 2007;13:315–323.
26. Lymperopoulos A, Rengo G, Gao E, Ebert SN, Dorn 2nd GW, Koch WJ. Reduction of sympathetic activity via adrenal-targeted GRK2 gene deletion attenuates heart failure progression and improves cardiac function after myocardial infarction. *J Biol Chem.* 2010;285:16378–16386.
27. Lymperopoulos A, Rengo G, Zincarelli C, Soltys S, Koch WJ. Modulation of adrenal catecholamine secretion by in vivo gene transfer and manipulation of G protein-coupled receptor kinase-2 activity. *Mol Ther.* 2008;16:302–307.
28. Lymperopoulos A, Rengo G, Koch WJ. Adrenal adrenoceptors in heart failure: fine-tuning cardiac stimulation. *Trends Mol Med.* 2007;13:503–511.
29. Conner DA, Mathier MA, Mortensen RM, et al. Beta-arrestin1 knockout mice appear normal but demonstrate altered cardiac responses to beta-adrenergic stimulation. *Circ Res.* 1997;81:1021–1026.
30. Bathgate-Siryk A, Dabul S, Pandya K, et al. Negative impact of β-arrestin-1 on post-myocardial infarction heart failure via cardiac and adrenal-dependent neurohormonal mechanisms. *Hypertension.* 2014;63:404–412.
31. McCrink KA, Maning J, Vu A, et al. β-Arrestin2 improves post-myocardial infarction heart failure via Sarco(endo)plasmic reticulum Ca_{2+}-ATPase-dependent positive Inotropy in cardiomyocytes. *Hypertension.* 2017;70:972–981.
32. McCrink KA, Maning J, Vu A, et al. Cardiac βarrestin2 improves contractility and adverse remodeling in heart failure, but is underexpressed in humans. *J Am Coll Cardiol* 2017;70:2948–2949.
33. McCrink KA, Brill A, Jafferjee M, et al. β_1-adrenoceptor Arg389Gly polymorphism confers differential β-arrestin-binding tropism in cardiac myocytes. *Pharmacogenomics.* 2016;17:1611–1620.

34. Parker BM, KA MC, Brill A, et al. Carvedilol uniquely stimulates βarrestin2-dependent SERCA2a activity via the β1-adrenergic receptor to exert positive inotropy in cardiac myocytes. *FASEB J.* 2018;32(suppl #1). Abstract #839.12.
35. Watari K, Nakaya M, Nishida M, Kim KM, Kurose H. β-arrestin2 in infiltrated macrophages inhibits excessive inflammation after myocardial infarction. *PLoS One.* 2013;8:e68351.
36. Noma T, Lemaire A, Naga Prasad SV, et al. Beta-arrestin-mediated beta1-adrenergic receptor transactivation of the EGFR confers cardioprotection. *J Clin Invest.* 2007;117:2445–2458.
37. Rajagopal K, Whalen EJ, Violin JD, et al. Beta-arrestin2-mediated inotropic effects of the angiotensin II type 1A receptor in isolated cardiac myocytes. *Proc Natl Acad Sci USA.* 2006;103:16284–16289.
38. Zhai P, Yamamoto M, Galeotti J, et al. Cardiac-specific overexpression of AT1 receptor mutant lacking G alpha q/G alpha i coupling causes hypertrophy and bradycardia in transgenic mice. *J Clin Invest.* 2005;115:3045–3056.
39. Ryba DM, Li J, Cowan CL, Russell B, Wolska BM, Solaro RJ. Long-term biased β-Arrestin signaling improves cardiac structure and function in dilated cardiomyopathy. *Circulation.* 2017;135:1056–1070.
40. Kim J, Zhang L, Peppel K, et al. Beta-arrestins regulate atherosclerosis and neointimal hyperplasia by controlling smooth muscle cell proliferation and migration. *Circ Res.* 2008;103:70–79.
41. Kim J, Ahn S, Rajagopal K, Lefkowitz RJ. Independent beta-arrestin2 and Gq/protein kinase Czeta pathways for ERK stimulated by angiotensin type 1A receptors in vascular smooth muscle cells converge on transactivation of the epidermal growth factor receptor. *J Biol Chem.* 2009;284:11953–11962.
42. Charles R, Namkung Y, Cotton M, Laporte SA, Claing A. β-Arrestin-mediated angiotensin II signaling controls the activation of ARF6 protein and endocytosis in migration of vascular smooth muscle cells. *J Biol Chem.* 2016;291:3967–3981.
43. Morinelli TA, Walker LP, Velez JC, Ullian ME. Clathrin-dependent internalization of the angiotensin II AT_1A receptor links receptor internalization to COX-2 protein expression in rat aortic vascular smooth muscle cells. *Eur J Pharmacol.* 2015;748:143–148.
44. Morinelli TA, Lee MH, Kendall RT, Luttrell LM, Walker LP, Ullian ME. Angiotensin II activates NF-κB through AT1A receptor recruitment of β-arrestin in cultured rat vascular smooth muscle cells. *Am J Physiol Cell Physiol.* 2013;304:C1176–C1186.
45. Lymperopoulos A, Rengo G, Zincarelli C, Kim J, Soltys S, Koch WJ. An adrenal beta-arrestin 1-mediated signaling pathway underlies angiotensin II-induced aldosterone production in vitro and in vivo. *Proc Natl Acad Sci USA.* 2009;106:5825–5830.
46. Lymperopoulos A, Rengo G, Zincarelli C, Kim J, Koch WJ. Adrenal beta-arrestin 1 inhibition in vivo attenuates post-myocardial infarction progression to heart failure and adverse remodeling via reduction of circulating aldosterone levels. *J Am Coll Cardiol.* 2011;57:356–365.
47. Lymperopoulos A, Sturchler E, Bathgate-Siryk A, et al. Different potencies of angiotensin receptor blockers at suppressing adrenal β-arrestin1-dependent post-myocardial infarction hyperaldosteronism. *J Am Coll Cardiol.* 2014;64:2805–2806.
48. Dabul S, Bathgate-Siryk A, Valero TR, et al. Suppression of adrenal βarrestin1-dependent aldosterone production by ARBs: head-to-head comparison. *Sci Rep.* 2015;5:8116.
49. Morris GE, Nelson CP, Brighton PJ, Standen NB, Challiss RA, Willets JM. Arrestins 2 and 3 differentially regulate ETA and P2Y2 receptor-mediated cell signaling and migration in arterial smooth muscle. *Am J Physiol Cell Physiol.* 2012;302:C723–C734.
50. Chen Y, Xu F, Zhang L, et al. GRK2/β-arrestin mediates arginine vasopressin-induced cardiac fibroblast proliferation. *Clin Exp Pharmacol Physiol.* 2017;44:285–293.

51. Barak LS, Oakley RH, Laporte SA, Caron MG. Constitutive arrestin-mediated desensitization of a human vasopressin receptor mutant associated with nephrogenic diabetes insipidus. *Proc Natl Acad Sci USA*. 2001;98:93–98.
52. Rochdi MD, Vargas GA, Carpentier E, et al. Functional characterization of vasopressin type 2 receptor substitutions (R137H/C/L) leading to nephrogenic diabetes insipidus and nephrogenic syndrome of inappropriate antidiuresis: implications for treatments. *Mol Pharmacol*. 2010;77:836–845.
53. Oligny-Longpré G, Corbani M, Zhou J, Hogue M, Guillon G, Bouvier M. Engagement of β-arrestin by transactivated insulin-like growth factor receptor is needed for V2 vasopressin receptor-stimulated ERK1/2 activation. *Proc Natl Acad Sci USA*. 2012;109: E1028–E1037.
54. Fuentes LQ, Reyes CE, Sarmiento JM, et al. Vasopressin up-regulates the expression of growth-related immediate-early genes via two distinct EGF receptor transactivation pathways. *Cell Signal*. 2008;20:1642–1650.
55. Albee LJ, LaPorte HM, Gao X, et al. Identification and functional characterization of arginine vasopressin receptor 1A: atypical chemokine receptor 3 heteromers in vascular smooth muscle. *Open Biol*. 2018;8:1.
56. Walters RW, Shukla AK, Kovacs JJ, et al. Beta-arrestin1 mediates nicotinic acid-induced flushing, but not its antilipolytic effect, in mice. *J Clin Invest*. 2009;119: 1312–1321.
57. Schaff M, Receveur N, Bourdon C, et al. β-arrestin-1 participates in thrombosis and regulates integrin aIIbβ3 signalling without affecting P2Y receptors desensitisation and function. *Thromb Haemost*. 2012;107:735–748.
58. Smith TH, Li JG, Dores MR, Trejo J. Protease-activated receptor-4 and purinergic receptor P2Y12 dimerize, co-internalize, and activate Akt signaling via endosomal recruitment of β-arrestin. *J Biol Chem*. 2017;292:13867–13878.
59. Morris GE, Nelson CP, Everitt D, et al. G protein-coupled receptor kinase 2 and arrestin2 regulate arterial smooth muscle P2Y-purinoceptor signalling. *Cardiovasc Res*. 2011;89:193–203.
60. Willets JM, Nash CA, Rainbow RD, Nelson CP, Challiss RA. Defining the roles of arrestin2 and arrestin3 in vasoconstrictor receptor desensitization in hypertension. *Am J Physiol Cell Physiol*. 2015;309:C179–C189.
61. Li D, D'Angelo L, Chavez M, Woulfe DS. Arrestin-2 differentially regulates PAR4 and ADP receptor signaling in platelets. *J Biol Chem*. 2011;286:3805–3814.
62. Khan A, Li D, Ibrahim S, Smyth E, Woulfe DS. The physical association of the P2Y12 receptor with PAR4 regulates arrestin-mediated Akt activation. *Mol Pharmacol*. 2014;86:1–11.
63. Roy RV, Ardeshirylajimi A, Dinarvand P, Yang L, Rezaie AR. Occupancy of human EPCR by protein C induces β-arrestin-2 biased PAR1 signaling by both APC and thrombin. *Blood*. 2016;128:1884–1893.
64. Yang P, Maguire J, Davenport AP. Apelin, Elabela/Toddler, biased agonists as novel therapeutic agents in the cardiovascular system. *Trends Pharmacol Sci*. 2015; 36:560–567.
65. Dai T, Ramirez-Correa G, Gao WD. Apelin increases contractility in failing cardiac muscle. *Eur J Pharmacol*. 2006;553:222–228.
66. Galvani S, Sanson M, Blaho VA, et al. HDL-bound sphingosine 1-phosphate acts as a biased agonist for the endothelial cell receptor S1P1 to limit vascular inflammation. *Sci Signal*. 2015;8:ra79.
67. Kho C, Lee A, Jeong D, et al. SUMO1-dependent modulation of SERCA2a in heart failure. *Nature*. 2011;477:601–605.
68. Kim IM, Tilley DG, Chen J, et al. Beta-blockers alprenolol and carvedilol stimulate beta-arrestin-mediated EGFR transactivation. *Proc Natl Acad Sci USA*. 2008;105: 14555–14560.

69. Wisler JW, DeWire SM, Whalen EJ, et al. A unique mechanism of beta-blocker action: carvedilol stimulates beta-arrestin signaling. *Proc Natl Acad Sci USA*. 2007;104:16657–16662.
70. Kaur K, Parra S, Chen R, et al. Gαi2 signaling: friend or foe in cardiac injury and heart failure? *Naunyn Schmiedebergs Arch Pharmacol*. 2012;385:443–453.
71. Nakaya M, Chikura S, Watari K, et al. Induction of cardiac fibrosis by β-blocker in G protein-independent and G protein-coupled receptor kinase 5/β-arrestin2-dependent signaling pathways. *J Biol Chem*. 2012;287:35669–35677.
72. Carr 3rd R, Schilling J, Song J, et al. β-Arrestin-biased signaling through the β2-adrenergic receptor promotes cardiomyocyte contraction. *Proc Natl Acad Sci USA*. 2016;113:E4107–E4116.
73. Littmann T, Göttle M, Reinartz MT, et al. Recruitment of β-arrestin 1 and 2 to the β2-adrenoceptor: analysis of 65 ligands. *J Pharmacol Exp Ther*. 2015;355:183–190.
74. Shukla AK, Westfield GH, Xiao K, et al. Visualization of arrestin recruitment by a G-protein-coupled receptor. *Nature*. 2014;512:218–222.
75. O'Hayre M, Eichel K, Avino S, et al. Genetic evidence that β-arrestins are dispensable for the initiation of β_2-adrenergic receptor signaling to ERK. *Sci Signal*. 2017;10(484).
76. Wang J, Hanada K, Staus DP, et al. Gαi is required for carvedilol-induced β1 adrenergic receptor β-arrestin biased signaling. *Nat Commun*. 2017;8:1706.
77. Gros R, Benovic JL, Tan CM, Feldman RD. G-protein-coupled receptor kinase activity is increased in hypertension. *J Clin Invest*. 1997;99:2087–2093.
78. Belmonte SL, Blaxall BC. G protein coupled receptor kinases as therapeutic targets in cardiovascular disease. *Circ Res*. 2011;109:309–319.
79. Lymperopoulos A, Brill A, McCrink KA. GPCRs of adrenal chromaffin cells & catecholamines: the plot thickens. *Int J Biochem Cell Biol*. 2016;77:213–219.
80. Jafferjee M, Reyes Valero T, Marrero C, McCrink KA, Brill A, Lymperopoulos A. GRK2 up-regulation creates a positive feedback loop for catecholamine production in chromaffin cells. *Mol Endocrinol*. 2016;30:372–381.
81. Cannavo A, Liccardo D, Lymperopoulos A, et al. GRK2 regulates α_2-adrenergic receptor-dependent catecholamine release in human adrenal chromaffin cells. *J Am Coll Cardiol*. 2017;69:1515–1517.
82. Maning J, Negussie S, Clark MA, Lymperopoulos A. Biased agonism/antagonism at the AngII-AT1 receptor: implications for adrenal aldosterone production and cardiovascular therapy. *Pharmacol Res*. 2017;125:14–20.
83. Lymperopoulos A. Beta-arrestin biased agonism/antagonism at cardiovascular seven transmembrane-spanning receptors. *Curr Pharm Des*. 2012;18:192–198.
84. Thomas WG, Thekkumkara TJ, Baker KM. Cardiac effects of AII. AT1A receptor signaling, desensitization, and internalization. *Adv Exp Med Biol*. 1996;396:59–69.
85. Saulière A, Bellot M, Paris H, et al. Deciphering biased-agonism complexity reveals a new active AT1 receptor entity. *Nat Chem Biol*. 2012;8:622–630.
86. Zidar DA, Violin JD, Whalen EJ, Lefkowitz RJ. Selective engagement of G protein coupled receptor kinases (GRKs) encodes distinct functions of biased ligands. *Proc Natl Acad Sci USA*. 2009;106:9649–9654.
87. Aplin M, Christensen GL, Schneider M, et al. Differential extracellular signal-regulated kinases 1 and 2 activation by the angiotensin type 1 receptor supports distinct phenotypes of cardiac myocytes. *Basic Clin Pharmacol Toxicol*. 2007;100:296–301.
88. Violin JD, DeWire SM, Yamashita D, et al. Selectively engaging β-arrestins at the angiotensin II type 1 receptor reduces blood pressure and increases cardiac performance. *J Pharmacol Exp Ther*. 2010;335:572–579.
89. Boerrigter G, Lark MW, Whalen EJ, Soergel DG, Violin JD, Burnett Jr JC. Cardiorenal actions of TRV120027, a novel ß-arrestin-biased ligand at the angiotensin II type

I receptor, in healthy and heart failure canines: a novel therapeutic strategy for acute heart failure. *Circ Heart Fail.* 2011;4:770–778.
90. Boerrigter G, Soergel DG, Violin JD, Lark MW, Burnett JC. TRV120027, a novel beta-arrestin biased ligand at the angiotensin II type I receptor, unloads the heart and maintains renal function when added to furosemide in experimental heart failure. *Circ Heart Fail.* 2012;5:627–634.
91. Rakesh K, Yoo B, Kim IM, Salazar N, Kim KS, Rockman HA. Beta-Arrestin-biased agonism of the angiotensin receptor induced by mechanical stress. *Sci Signal.* 2010;3: ra46.
92. Abraham DM, Davis 3rd RT, Warren CM, et al. β-Arrestin mediates the Frank-Starling mechanism of cardiac contractility. *Proc Natl Acad Sci USA.* 2016;113:14426–14431.
93. MacKenzie A. Endothelium-derived vasoactive agents, AT1 receptors and inflammation. *Pharmacol Ther.* 2011;131:187–203.
94. Ahn S, Kim J, Hara MR, Ren XR, Lefkowitz RJ. Beta-arrestin-2 mediates anti-apoptotic signaling through regulation of BAD phosphorylation. *J Biol Chem.* 2009;284:8855–8865.
95. Kendall RT, Lee MH, Pleasant DL, et al. Arrestin-dependent angiotensin AT1 receptor signaling regulates Akt and mTor-mediated protein synthesis. *J Biol Chem.* 2014;289:26155–26166.
96. Wilson PC, Lee MH, Appleton KM, et al. The arrestin-selective angiotensin AT1 receptor agonist [Sar1,Ile4,Ile8]-AngII negatively regulates bradykinin B2 receptor signaling via AT1-B2 receptor heterodimers. *J Biol Chem.* 2013;288:18872–18884.
97. Liu Y, Wang Z, Li J, et al. Inhibition of 5-Hydroxytryptamine receptor 2B reduced vascular restenosis and mitigated the β-arrestin2-mammalian target of rapamycin/p70S6K pathway. *Am Heart Assoc.* 2018;7:3.
98. Nishimura A, Sunggip C, Tozaki-Saitoh H, et al. Purinergic P2Y6 receptors heterodimerize with angiotensin AT1 receptors to promote angiotensin II-induced hypertension. *Sci Signal.* 2016;9:ra7.
99. Jean-Charles PY, Zhang L, Wu JH, et al. Ubiquitin-specific protease 20 regulates the reciprocal functions of β-arrestin2 in toll-like receptor 4-promoted nuclear factor κB (NFκB) activation. *J Biol Chem.* 2016;291:7450–7464.
100. Lymperopoulos A, Aukszi B. Angiotensin receptor blocker drugs and inhibition of adrenal beta-arrestin-1-dependent aldosterone production: implications for heart failure therapy. *World J Cardiol.* 2017;9:200–206.
101. Tóth AD, Prokop S, Gyombolai P, et al. Heterologous phosphorylation-induced formation of a stability lock permits regulation of inactive receptors by β-arrestins. *J Biol Chem.* 2018;293:876–892.
102. Balakumar P, Jagadeesh G. Structural determinants for binding, activation, and functional selectivity of the angiotensin AT1 receptor. *J Mol Endocrinol.* 2014;53:R71–R92.
103. Nikiforovich GV, Zhang M, Yang Q, et al. Interactions between conserved residues in transmembrane helices 2 and 7 during angiotensin AT1 receptor activation. *Chem Biol Drug Des.* 2006;68:239–249.
104. Valero TR, Sturchler E, Jafferjee M, et al. Structure-activity relationship study of angiotensin II analogs in terms of β-arrestin-dependent signaling to aldosterone production. *Pharmacol Res Perspect.* 2016;4:e00226.
105. Liu CH, Gong Z, Liang ZL, et al. Arrestin-biased AT1R agonism induces acute catecholamine secretion through TRPC3 coupling. *Nat Commun.* 2017;8:14335.
106. Ivey ME, Osman N, Little PJ. Endothelin-1 signalling in vascular smooth muscle: pathways controlling cellular functions associated with atherosclerosis. *Atherosclerosis.* 2008;199:237–247.

107. Lozić M, Šarenac O, Murphy D, Japundžić-Žigon N. Vasopressin, central autonomic control and blood pressure regulation. *Curr Hypertens Rep*. 2018;20:11.
108. Oakley RH, Laporte SA, Holt JA, Caron MG, Barak LS. Differential affinities of visual arrestin, beta arrestin1, and beta arrestin2 for G protein-coupled receptors delineate two major classes of receptors. *J Biol Chem*. 2000;275:17201–17210.
109. Digby JE, Ruparelia N, Choudhury RP. Niacin in cardiovascular disease: recent preclinical and clinical developments. *Arterioscler Thromb Vasc Biol*. 2012;32:582–588.
110. Pike NB. Flushing out the role of GPR109A (HM74A) in the clinical efficacy of nicotinic acid. *J Clin Invest*. 2005;115:3400–3403.
111. Kim HY, Jadhav VB, Jeong DY, et al. Discovery of 4-(phenyl)thio-1H-pyrazole derivatives as agonists of GPR109A, a high affinity niacin receptor. *Arch Pharm Res*. 2015;38:1019–1032.
112. Gachet C. P2 receptors, platelet function and pharmacological implications. *Thromb Haemost*. 2008;99:466–472.
113. Hardy AR, Conley PB, Luo J, Benovic JL, Poole AW, Mundell SJ. P2Y1 and P2Y12 receptors for ADP desensitize by distinct kinase-dependent mechanisms. *Blood*. 2005;105:3552–3560.
114. Soh UJ, Dores MR, Chen B, Trejo J. Signal transduction by protease-activated receptors. *Br J Pharmacol*. 2010;160:191–203.
115. Mosnier LO, Sinha RK, Burnier L, Bouwens EA, Griffin JH. Biased agonism of protease-activated receptor 1 by activated protein C caused by noncanonical cleavage at Arg46. *Blood*. 2012;120:5237–5246.
116. Soh UJ, Trejo J. Activated protein C promotes protease-activated receptor-1 cytoprotective signaling through β-arrestin and dishevelled-2 scaffolds. *Proc Natl Acad Sci USA*. 2011;108:E1372–E1380.
117. Tatemoto K, Hosoya M, Habata Y, et al. Isolation and characterization of a novel endogenous peptide ligand for the human APJ receptor. *Biochem Biophys Res Commun*. 1998;251:471–476.
118. Iturrioz X, Gerbier R, Leroux V, Alvear-Perez R, Maigret B, Llorens-Cortes C. By interacting with the C-terminal Phe of apelin, Phe255 and Trp259 in helix VI of the apelin receptor are critical for internalization. *J Biol Chem*. 2010;285:32627–32637.
119. Iturrioz X, Alvear-Perez R, De Mota N, et al. Identification and pharmacological properties of E339-3D6, the first nonpeptidic apelin receptor agonist. *FASEB J*. 2010;24:1506–1517.
120. Brame AL, Maguire JJ, Yang P, et al. Design, characterization, and first-in-human study of the vascular actions of a novel biased apelin receptor agonist. *Hypertension*. 2015;65:834–840.
121. Murza A, Sainsily X, Côté J, et al. Structure-activity relationship of novel macrocyclic biased apelin receptor agonists. *Org Biomol Chem*. 2017;15:449–458.
122. Pyne NJ, Pyne S. Sphingosine 1-phosphate receptor 1 signaling in mammalian cells. *Molecules*. 2017;22:3.
123. Pearson TA, Mensah GA, Alexander RW, Anderson JL, Cannonal RO. Markers of inflammation and cardiovascular disease: application to clinical and public health practice: a statement for healthcare professionals from the centers for disease control and prevention and the American heart association. *Circulation*. 2003;107:499–511.
124. Willerson JT, Ridker PM. Inflammation as a cardiovascular risk factor. *Circulation*. 2004;109:102–110.
125. Hughes JE, Srinivasan S, Lynch KR, Proia RL, Ferdek P, Hedrick CC. Sphingosine-1-phosphate induces an antiinflammatory phenotype in macrophages. *Circ Res*. 2008;102:950–958.
126. Srivastava A, Gupta B, Gupta C, Shukla AK. Emerging functional divergence of β-arrestin isoforms in GPCR function. *Trends Endocrinol Metab*. 2015;26:628–642.

127. Lymperopoulos A, Negussie S. βArrestins in cardiac G protein-coupled receptor signaling and function: partners in crime or "good cop, bad cop"? *Int J Mol Sci.* 2013;14:24726–24741.
128. Zhan X, Gimenez LE, Gurevich VV, Spiller BW. Crystal structure of arrestin-3 reveals the basis of the difference in receptor binding between two non-visual subtypes. *J Mol Biol.* 2011;406:467–478.
129. Zhuo Y, Vishnivetskiy SA, Zhan X, Gurevich VV, Klug CS. Identification of receptor binding-induced conformational changes in non-visual arrestins. *J Biol Chem.* 2014;289:20991–21002.
130. Gurevich VV, Gurevich EV. Extensive shape shifting underlies functional versatility of arrestins. *Curr Opin Cell Biol.* 2014;27:1–9.
131. Xiao K, McClatchy DB, Shukla AK, et al. Functional specialization of beta-arrestin interactions revealed by proteomic analysis. *Proc Natl Acad Sci USA.* 2007;104:12011–12016.

CHAPTER THREE

Global Aquatic Hazard Assessment of Ciprofloxacin: Exceedances of Antibiotic Resistance Development and Ecotoxicological Thresholds

Kaitlyn R. Kelly*, Bryan W. Brooks*,†,1
*Environmental Health Science Program, Department of Environmental Science, Baylor University, Waco, TX, United States
†Institute of Biomedical Studies, Baylor University, Waco, TX, United States
[1]Corresponding author: e-mail address: bryan_brooks@Baylor.edu

Contents

1. Background 60
2. Materials and Methods 62
   2.1 Literature Review 62
   2.2 Probabilistic Aquatic Hazard Assessments 62
3. Results and Discussion 63
   3.1 Ciprofloxacin in Municipal and Hospital Sewage and Effluent Discharges 64
   3.2 Ciprofloxacin in Freshwater, Marine Systems and Groundwater 72
4. Conclusions 74
References 75

Abstract

Rapid urbanization represents a global megatrend that is resulting in an increasingly urban water cycle frequently contaminated by human medicines and other chemicals. Concentration of chemical consumption is occurring faster than implementation of environmental health systems and interventions, particularly in megacities of developing countries, while 80% of global sewage production remains untreated. In these urban areas, antibiotics in the environment influence development of antibiotic resistance (ABR) by pathogens. Here, we examined the occurrence of ciprofloxacin, designated a critically important antibiotic by the World Health Organization, in fresh surface water, groundwater, saltwater, treated municipal effluent, raw municipal sewage, treated hospital effluent, and raw hospital sewage around the world. We examined 260 articles reporting ciprofloxacin in these water matrices, including 111 studies from Europe, 105 from Asia, and 32 from North America. Only five and seven studies were identified from South America and Africa, respectively. Probabilistic hazard assessments were then

Progress in Molecular Biology and Translational Science, Volume 159
ISSN 1877-1173
https://doi.org/10.1016/bs.pmbts.2018.07.004

performed to identify probabilities of exceedances of reported ABR and ecotoxicological predicted no effect concentrations (PNEC) for ciprofloxacin in water. Remarkably, across all geographic regions, 58% of municipal effluents exceeded the ABR PNEC (100 ng/L) for ciprofloxacin, whereas ecotoxicity PNEC (1200 ng/L) was exceeded 16% of the time. In surface water, 25% reported concentrations exceeded the ABR PNEC with 31% and 29% in Europe and North America, respectively, compared to 67% exceedance identified for Africa. Such observations highlight the importance of global scanning assessments to identify matrices and regions, including areas strategically advancing water reuse, for additional study and to prioritize locations for interventions aimed at reducing ABR development.

1. BACKGROUND

Increasing population growth and urbanization is resulting in an increasingly urban water cycle that is frequently contaminated by human medicines. By 2050, 70% of people will live in urban areas with 50% living in the Asia-Pacific by 2018, yet 80% of the global sewage production is released as untreated waste to the environment.[1] Though wastewater effluents routinely represent the primary pathway for introduction of pharmaceuticals to environment, most wastewater treatment plants (WWTP) technologies were not designed to specifically treat pharmaceuticals and other contaminants of emerging concern.[1] Unfortunately, pharmaceutical occurrence and risks in marine systems[2] and developing countries[3] have received comparatively less attention. However, understanding risks of pharmaceuticals in the environment was recently reported as a priority research need for environmental quality in Latin America.[4] Identifying hotspots, or regions of elevated environmental contamination, for pharmaceuticals[5,6] and other contaminants[7] have further been identified as key research questions during international horizon scanning efforts. For example, municipal effluent dominated and dependent systems, which are surface waters with the majority of instream flows from water reclamation or raw sewage discharges, often represent worse case scenarios for exposure to pharmaceuticals, consumer chemicals and endogenous compounds in developed countries.[8,9] These aquatic systems have been identified as important management units in the face of climate change.[10]

Antibiotics represent particularly important classes of pharmaceuticals in the environment because they may directly result in adverse outcomes in ecosystems[11,12] and indirectly present risks to human health when they influence the development of antibiotic resistance (ABR) in pathogens.[5]

Antimicrobial resistance accounts for 700,000 deaths per year,[13] and subsequently is now recognized as a major global health threat.[14] The presence of antibiotics in the environment exerts strong selective pressure for development of ABR in pathogens.[15] This threat decidedly requires an advanced understanding of urbanization, environment and health dimensions of ABR by examining the life cycle of antibiotics, including wastewater treatment.[16] For example, alarmingly elevated levels of pharmaceuticals, including the antibiotic ciprofloxacin, were reported by Larsson et al.[17] from pharmaceutical manufacturing facilities in India, which resulted in high levels of resistance and gene transfer elements.[18] It is therefore important to note that ABR development and antibiotic residues in the environment have been identified as priority questions to understand environmental risks of pharmaceuticals.[5,6]

The World Health Organization has identified the fluoroquinolone class of antibiotics, including ciprofloxacin, as critically important antibiotics for management of ABR microorganisms.[19] Ciprofloxacin, which targets both gram-negative and gram-positive bacteria in human and veterinary medicine,[20] is often used as a drug of last resort for severe bacterial infections and is thus one of the most important antibiotics. Ciprofloxacin was first discovered in the 1980s and was developed after finding that replacing the ethyl group with a cyclopropyl group increased its broad-spectrum capabilities.[21] Ciprofloxacin specifically targets type II topoisomerase DNA gyrase and topoisomerase IV, disrupting DNA replication, transcription, repair, and recombination in bacteria.[22] Oral doses are readily absorbed in the gastrointestinal system and then widely distributed through the body.[23] The plasma half-life is 4 h after peak levels at 1–2 h after dosing and it is largely excreted unchanged (40–50%) in the urine.[23] Ciprofloxacin is routinely observed in municipal sewage and wastewater treatment plants discharges and is not considered readily biodegradable.[24] ABR against ciprofloxacin and other fluoroquinolones results from mutation in type II topoisomerase encoding genes, drug efflux, and membrane permeability.[25] Ciprofloxacin and other antibiotics can be introduced to surface waters through other sources, including sewage sludge application to terrestrial systems and from landfill leachate. For example, we recently observed ciprofloxacin levels in leachate from active and closed landfills in Hong Kong[26] that exceeded thresholds for ABR development.

Whether such observations of ABR threshold exceedances by ciprofloxacin extend to other regions or aquatic systems requires attention. Herein, global scanning approaches employing probabilistic hazard assessments provide unique opportunities to identify global hot spots requiring environmental management,[27–29] particularly for antibiotics and ABR.[16] Therefore, the primary objective of this study was to examine the occurrence of ciprofloxacin in aquatic matrices, including inland waters and marine systems, groundwater, and municipal and hospital wastewater (treated effluent and raw sewage). We then performed probabilistic aquatic hazard assessments for ciprofloxacin to identify potential exceedances of predicted no effect concentrations (PNEC) for ABR and ecotoxicity.

2. MATERIALS AND METHODS

2.1 Literature Review

A literature search was conducted for studies observing ciprofloxacin in inland waters, groundwater, coastal and marine systems, municipal wastewater (treated effluent and raw sewage), and hospital wastewater (treated effluent and raw sewage). Studies quantifying ciprofloxacin in industrial effluent and veterinary effluent were specifically excluded. Search terms ciprofloxacin and occurrence and water or surface or effluent -clinical -oral -mesocosms in Google Scholar resulted in 6954 hits between 1990 and April 2018. Of those hits, 345 publications were relevant with 210 reporting environmental data. An additional literature search was subsequently conducted with Scopus using the search terms ciprofloxacin and surface water, ciprofloxacin and sewage or effluent, ciprofloxacin and groundwater, ciprofloxacin and seawater, and ciprofloxacin and hospital wastewater. During the Scopus search, which was informed by the Google Scholar results, the terms clinical, oral, and mesocosms were excluded. This Scopus search resulted in 137 new relevant articles with 50 containing environmental data. These Google Scholar and Scopus results included 1094 detects of ciprofloxacin across all water matrices. Quantitative data on the concentration (e.g., mean, median, minimum, maximum) of ciprofloxacin in the designated water matrixes was collected from relevant manuscripts and collated. Major geographic location of environmental data (e.g., Africa, Asia-Pacific, Europe, North America, South America) collection was also recorded for ciprofloxacin.

2.2 Probabilistic Aquatic Hazard Assessments

The methods in the present study followed an approach described previously by our research group.[27–29,16] Probabilistic environmental exposure distributions (EED) for each water matrix were created using the maximum measured environmental concentration (MEC). EED graphs were created using SigmaPlot 11.0 (Systat Software, Inc. San Jose, CA, USA). The distributions were then used to perform probabilistic aquatic hazard assessments to determine probabilities of identifying ciprofloxacin at specific concentrations in each water matrix. The collected MECs were ranked using the Weibull formula Eq. (1) and assigned weighted percentiles:

$$j = (i \times 100)/(n + 1) \quad (1)$$

where j is the percent rank, i the MEC rank, n the number of chemicals, and $n+1$ takes into account that there is always one less than all measured occurrences.[19] Ciprofloxacin MECs were plotted on the x-axis (log common transformed) against the percent rank as the y-axis (probability transformed). Lines of linear regression were assigned and analyzed for their slope, and y-intercept values to calculate centile values using the equation:

$$\text{Centile value} = \text{NORMDIST}\left((m \times \log 10(x)) + b\right) \quad (2)$$

where NORMDIST is a standard normal cumulative distribution function, x the threshold value, m the slope and b the y-intercept from the linear regressions.[29] Probabilistic ciprofloxacin distribution values were calculated for the following centiles: 1, 5, 25, 50, 75, 95, and 99 using Microsoft Excel (Microsoft Excel, 2010 Microsoft Corp, Redmond, WA, USA). Whereas a PNEC value of 100 ng/L[30] was employed for an ABR threshold, a PNEC of 1200 ng/L was used for ciprofloxacin ecotoxicity.[15] When data were sufficient to develop probability distributions ($n \geq 5$), PEHAs were performed for each geographic region and water matrix.

3. RESULTS AND DISCUSSION

The number of publications reporting ciprofloxacin in wastewater effluent, surface waters, and groundwater has steadily increased in the last decade (Fig. 1). This pattern has liked resulted because of the increasing attention given to antibiotics and other pharmaceuticals in the environment[31] and the advancement and accessibility of advanced analytical chemistry methods and instrumentation.[16] Out of the 260 studies observing ciprofloxacin in water matrices, 83% occurred in Asia (105 studies) and Europe (111 studies). Only five studies in South America were identified, three of which reported ciprofloxacin in surface water, and two in raw hospital sewage. Such limited observations from South America and other large geographic regions (e.g., Africa) are consistent with our recent observations

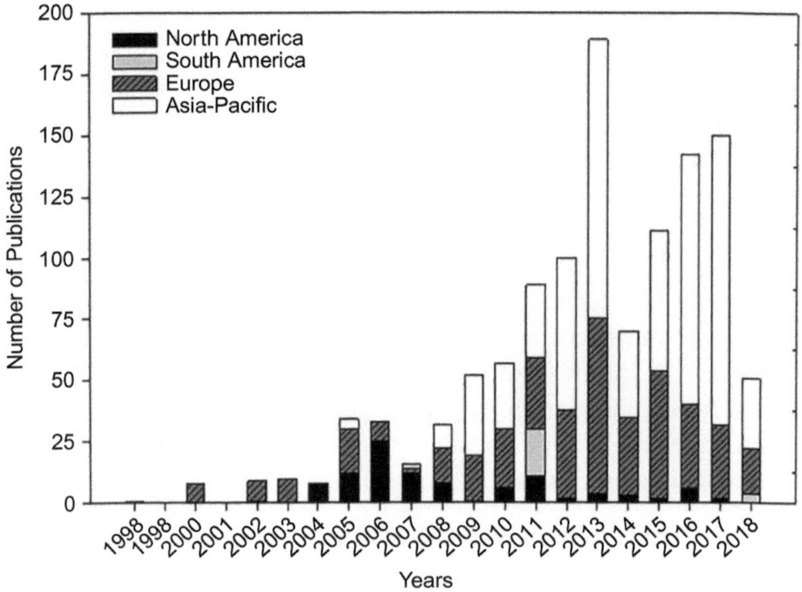

Fig. 1 Refereed studies identifying ciprofloxacin in various aquatic matrices through time (until May 2018).

of limited environmental monitoring data for erythromycin,[16] and highlight the importance of understanding intersections of increasingly concentrated antibiotic use in cities, environmental exposures and reservoirs for ABR development, and adverse health outcomes in rapidly growing and urbanizing regions of developing countries.

3.1 Ciprofloxacin in Municipal and Hospital Sewage and Effluent Discharges

We examined raw municipal sewage because of the global prevalence of improper sanitation and poorly implemented and maintained sewage treatment technologies. The maximum concentration of ciprofloxacin out of 31 detects reported in sewage was 341,000 ng/L (Table 1). PEHA were then conducted to examine levels of ciprofloxacin in municipal sewage, and to identify exceedance likelihoods of PNECs for ABR and ecotoxicity. In the present study, consistent with our previously reported methods, we focused on maximum reported levels of ciprofloxacin because minimum, median and mean values and raw data could not be consistently obtained across the published literature. Elevated levels of ciprofloxacin were observed in sewage with 50th and 95th centiles identified at 312

Table 1 Detection Frequency and Geographic Distribution of Ciprofloxacin Among Aquatic Matrices and Geographic Regions

	Detection (ng/L)			Geographic Distribution				
Matrix	**# Times Detected**	**Min**	**Max**	**Africa**	**Asia**	**Europe**	**North America**	**South America**
All	1094	ND	5,250,000	36	657	458	104	25
Inland waters	490	ND	245,550	14	294	149	17	16
Groundwater	111	ND	443	–	96	14	1	–
Marine systems	57	ND	9410	–	50	6	1	–
Treated municipal effluent	316	ND	5,250,000	17	81	168	50	–
Raw municipal sewage	32	<LOD	341,000	–	4	27	1	–
Treated hospital effluent	68	<LOQ	1,830,400	4	44	16	2	2
Raw hospital sewage	20	119	600,000	1	6	6	5	2

and 13,744 ng/L, respectively (Table 2). PNECs for ABR and ecotoxicity were specifically exceeded 87% and 19% of the time, respectively, in municipal sewage (Table 3; Fig. 2A).

Of the 1094 total detects of ciprofloxacin across water matrices, 316 were in treated municipal wastewater effluent (Table 1). Because WWTP effluent is a major source of antibiotics in the environment, most research has investigated ciprofloxacin in this matrix.[16] The first detection of ciprofloxacin in municipal wastewater effluent apparently was in the Glatt Valley Watershed in Switzerland in 2002.[32] Since this initial report, 168 other detects of ciprofloxacin in wastewater effluent were identified within Europe, followed by 81 from Asia-Pacific (Table 1). No detects of ciprofloxacin were identified in municipal wastewater effluent from South America; 17 detects were reported in Africa.

For wastewater effluent, ciprofloxacin observations were examined by geographic region. The highest number of observations were in Asia-Pacific (40), Europe (84), and North America[33] (Table 2). 50th centile levels across all geographic regions (188 ng/L; Table 2) were higher than the ABR PNEC of 100 ng/L with an 85% exceedance (Table 3; Fig. 2B) in this matrix. Exceedances of PNECs for ecotoxicity were also observed, though less frequently (15%; Table 3; Fig. 2B). Thus, these exceedances were comparable to but slightly lower than those observed for raw municipal sewage (Table 3). Differences were also observed across major geographic regions; for example, 50th centile values were slightly lower in Asia (154 ng/L) compared to Europe (233 ng/L) and North America (193 ng/L). Similarly, percent exceedances of the ABR PNEC were less likely to be observed in Asia (36%) compared to Europe (69%) and North America (68%) (Fig. 2B).

We then separately examined hospital raw sewage and effluent because of concentrated use of ciprofloxacin in hospital settings. Extremely high concentrations of ciprofloxacin have been reported in raw hospital sewage (600,000 ng/L) and treated hospital effluent (1,830,400 ng/L) (Table 1). A majority of the reported detects were in Asia, followed by Europe, though ciprofloxacin has been reported across all geographic regions in both raw and treated hospital effluent. Compared to raw municipal sewage and effluent, elevated levels of ciprofloxacin have been identified in hospital wastewater. For example, 50th centile values were 4137 ng/L and 3637 ng/L in raw sewage and effluents from hospitals, respectively (Table 2). It is particularly important to note that such observations exceed ABR PNECs by over an order of magnitude, and also exceed the PNEC for ecotoxicity (Table 3). In fact, ciprofloxacin PNECs for ABR and ecotoxicity were exceeded

Table 2 Equations for Regressions Lines and Values Corresponding to Various Centile Values for Environmental Exposure Distributions (EEDs) of Maximum Reported Measured Environmental Concentrations (MECs) for Ciprofloxacin (ng/L) in Effluent, Inland Waters, Groundwater and Marine Systems

						Centile Value (ng/L)							
Matrix	**Region**	***n***	**r^2**	**Slope**	**Intercept**	**1**	**5**	**10**	**20**	**50**	**75**	**95**	**99**
Inland waters	All regions	338	0.93	1.02	−1.67	0.23	1.05	2.39	6.48	43.42	199.53	1790.01	8356.87
	Africa	6	0.89	0.46	−1.17	0.00	0.09	0.56	5.14	354.14	10,525.93	1,385,286.06	42,650,686.44
	Asia	231	0.89	0.99	−1.59	0.18	0.88	2.04	5.67	40.20	193.01	1844.56	9002.58
	Europe	85	0.95	1.14	−1.98	0.50	1.96	4.08	9.93	54.32	212.05	1504.51	5956.94
	N. America	7	0.99	3.16	−5.79	12.53	20.60	26.85	37.02	68.41	111.91	227.17	373.51
	S. America	9	0.85	0.67	−0.70	0.00	0.04	0.14	0.61	10.98	111.18	3107.74	32,231.61
Marine systems	All regions	54	0.77	1.54	−2.08	0.69	1.92	3.31	6.40	22.59	62.12	266.12	739.37
	Asia	48	0.94	2.59	−3.19	2.16	3.96	5.48	8.10	17.12	31.20	73.96	135.60
	Europe	5	0.93	3.15	−4.95	6.77	11.14	14.52	20.03	37.02	60.58	123.04	202.38
Groundwater	All regions	85	0.81	1.69	−1.47	0.31	0.79	1.30	2.36	7.42	18.59	69.70	176.43
	Asia	79	0.88	2.53	−1.94	0.70	1.31	1.82	2.72	5.86	10.84	26.26	48.86
Treated municipal	All regions	212	0.95	0.94	−2.14	0.63	3.35	8.15	23.92	187.73	/	10,530.02	55,850.36
Effluent	Asia	59	0.81	0.65	−1.41	0.04	0.43	1.59	7.63	153.74	1705.47	54,365.56	618,342.84

Continued

Table 2 Equations for Regressions Lines and Values Corresponding to Various Centile Values for Environmental Exposure Distributions (EEDs) of Maximum Reported Measured Environmental Concentrations (MECs) for Ciprofloxacin (ng/L) in Effluent, Inland Waters, Groundwater and Marine Systems—cont'd

						Centile Value (ng/L)							
Matrix	**Region**	***n***	r^2	**Slope**	**Intercept**	**1**	**5**	**10**	**20**	**50**	**75**	**95**	**99**
	Europe	108	0.99	1.14	−2.70	2.12	8.41	17.52	42.60	233.08	909.99	6457.19	25,568.91
	N. America	41	0.89	0.99	−2.27	0.88	4.27	9.91	27.47	193.20	922.48	8744.73	42,437.84
Raw municipal	All regions	31	0.87	2.31	−3.68	3.84	7.56	10.86	16.83	38.91	76.16	200.16	394.52
Sewage	Europe	27	0.75	1.00	−2.49	1.47	7.06	16.30	44.87	311.52	1471.93	13,744.42	65,998.41
Treated hospital	All regions	26	0.98	0.61	−2.18	0.58	7.49	29.36	153.53	3636.57	45,943.92	1,765,910.62	22,905,756.43
Effluent	Asia	13	0.91	0.42	−1.34	0.00	0.18	1.36	15.29	1564.18	63,814.99	13,246,100.18	561,631,729.2
	Europe	9	0.96	1.64	−6.54	378.27	987.44	1646.87	3059.57	10,006.51	25,864.84	101,403.67	264,705.99
Raw hospital sewage	All regions	17	0.96	0.85	−3.08	7.69	48.51	129.51	425.31	4136.60	25,608.89	352,735.05	2,225,450.58
	Europe	5	0.81	0.73	−1.47	0.07	0.58	1.81	7.21	100.94	836.90	17,549.20	148,736.22
	N. America	5	0.98	0.87	−3.03	6.33	38.05	98.99	315.07	2886.38	17,032.19	218,957.52	1,316,079.76

For each distribution, '*n*' represents the number of MECs reported and used in each regression.

Table 3 Exceedances of Measured Environmental Concentrations of Ciprofloxacin Relative to Predicted No Effect Concentrations (PNEC) for Antibiotic Resistance (ABR) Development (100 ng/L) and Ecotoxicity (1200 ng/L)

Matrix	Region	Total Detects	Detects Over ABR PNEC of 100 ng/L	% Exceedance of ABR PNEC	Detects Over Ecotoxicity PNEC of 1200 ng/L	% Exceedance of Ecotoxicity PNEC
Inland waters	All regions	338	84	24.85	29	8.58
	Africa	6	4	66.67	2	33.33
	Asia	231	49	21.21	19	8.23
	Europe	85	26	30.59	8	9.41
	N. America	7	2	28.57	0	0.00
	S. America	9	3	33.33	0	0.00
Marine systems	All regions	54	1	1.85	1	1.85
	Asia	48	0	0.00	0	0.00
	Europe	5	0	0.00	0	0.00
	N. America	1	1	100.00	1	100.00
Groundwater	All regions	85	6	7.06	0	0.00
	Asia	79	2	2.53	0	0.00
	Europe	5	4	80.00	0	0.00
	N. America	1	0	0.00	0	0.00
Treated municipal	All regions	218	126	57.80	34	15.60
Effluent	Africa	4	3	75.00	2	50.00

Continued

Table 3 Exceedances of Measured Environmental Concentrations of Ciprofloxacin Relative to Predicted No Effect Concentrations (PNEC) for Antibiotic Resistance (ABR) Development (100 ng/L) and Ecotoxicity (1200 ng/L)—cont'd

Matrix	Region	Total Detects	Detects Over ABR PNEC of 100 ng/L	% Exceedance of ABR PNEC	Detects Over Ecotoxicity PNEC of 1200 ng/L	% Exceedance of Ecotoxicity PNEC
	Asia	59	21	35.59	9	15.25
	Europe	108	75	69.44	21	19.44
	N. America	41	28	68.29	3	7.32
Raw municipal	All regions	31	27	87.10	6	19.35
Sewage	Asia	4	2	50.00	2	50.00
	Europe	27	25	92.59	4	14.81
Treated hospital	All regions	26	22	84.62	17	65.38
Effluent	Asia	13	9	69.23	5	38.46
	Europe	9	9	100.00	9	100.00
	N. America	1	1	100.00	1	100.00
	S. America	2	2	100.00	2	100.00
Raw hospital	All regions	17	17	100.00	14	82.35
Sewage	Africa	1	1	100.00	1	100.00
	Asia	4	4	100.00	4	100.00
	Europe	5	5	100.00	3	60.00
	N. America	5	5	100.00	4	80.00
	S. America	2	2	100.00	2	100.00

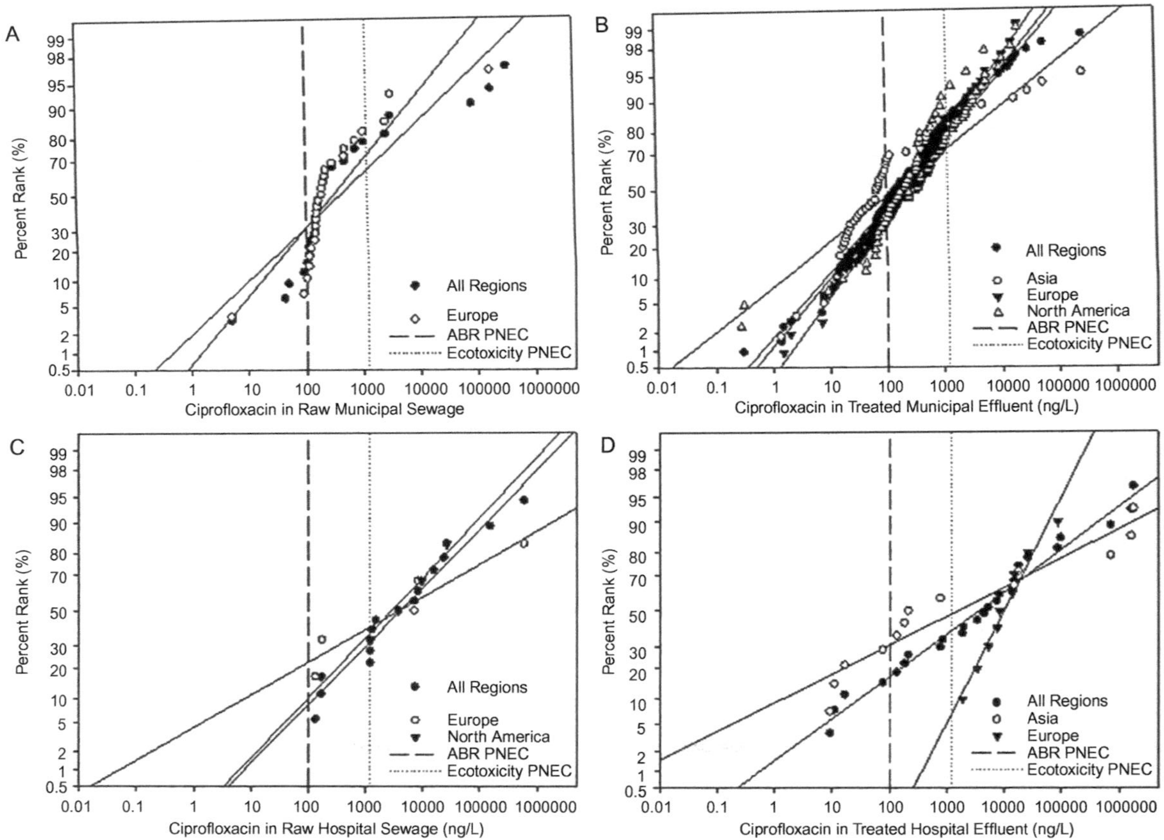

Fig. 2 Measured Environmental Concentration distributions of maximum reported ciprofloxacin in raw municipal sewage (A), municipal effluent (B), raw hospital sewage (C) and hospital effluent (D). Vertical lines correspond to Predicted No Effect Concentrations (PNECs) for antibiotic resistance (ABR) development (100 ng/L) and ecotoxicity (1200 ng/L).

100% of the time in raw hospital sewage (Table 3; Fig. 2C). Exceedances of the ABR and ecotoxicity PNECs were observed 85% and 65% of the time, respectively, in effluents from hospitals (Table 3; Fig. 2D). Such observations indicate the important contribution of hospital waste streams of antibiotics and suggest that industrial pretreatment of hospital sewage should be considered during future environmental management efforts, particularly in regions strategically employing water reuse.

3.2 Ciprofloxacin in Freshwater, Marine Systems and Groundwater

Inland freshwaters have also been a heavily researched water matrix, with 490 total detects of ciprofloxacin (Table 1). Inland waters have been most commonly investigated in Asia-Pacific (294 detects) and Europe (149 detects; Table 1). The highest reported concentration of ciprofloxacin in inland waters was 245,550 ng/L, which is over an order of magnitude lower than the maximum concentration reported from municipal effluent discharges (5,250,000 ng/L).

As performed for effluent discharges, we examined these observations by geographic region (Table 2). 50th Centile values for all observations (43 ng/L) and by geographic regions were typically much lower in inland surface water than effluent (Table 2), likely reflecting instream dilution of effluent discharges. Africa presented a noticeable exception with a 50th centile value (354 ng/L) exceeding the ABR PNEC, though relatively few data points were available from this region compared to Asia-Pacific or Europe. Similarly, percent exceedances of the ABR and ecotoxicity PNEC values in inland waters were higher in Africa than other geographic regions (Table 3; Fig. 3A). Across all regions, the ABR and ecotoxicity PNECs were exceeded 25% and 9% of the time, respectively (Table 3; Fig. 3A). Clearly, such elevated exceedance frequencies represent an important observation and deserve future study of the extent of ABR pathogen occurrence and ecological consequences associated with such elevated ciprofloxacin residues in urban surface waters.

In contrast to inland waters, marine and coastal systems have received much less study, with only 57 detections (Table 1). However, estuaries are some of the most productive systems, increasingly support aquaculture operations, and provide important ecosystem services.[33] Further, much of the global population, including an increasing percentage located in megacities, is located near coastlines and subsequently discharges wastewater of varying quality there. Additionally, elderly populations, a segment of the

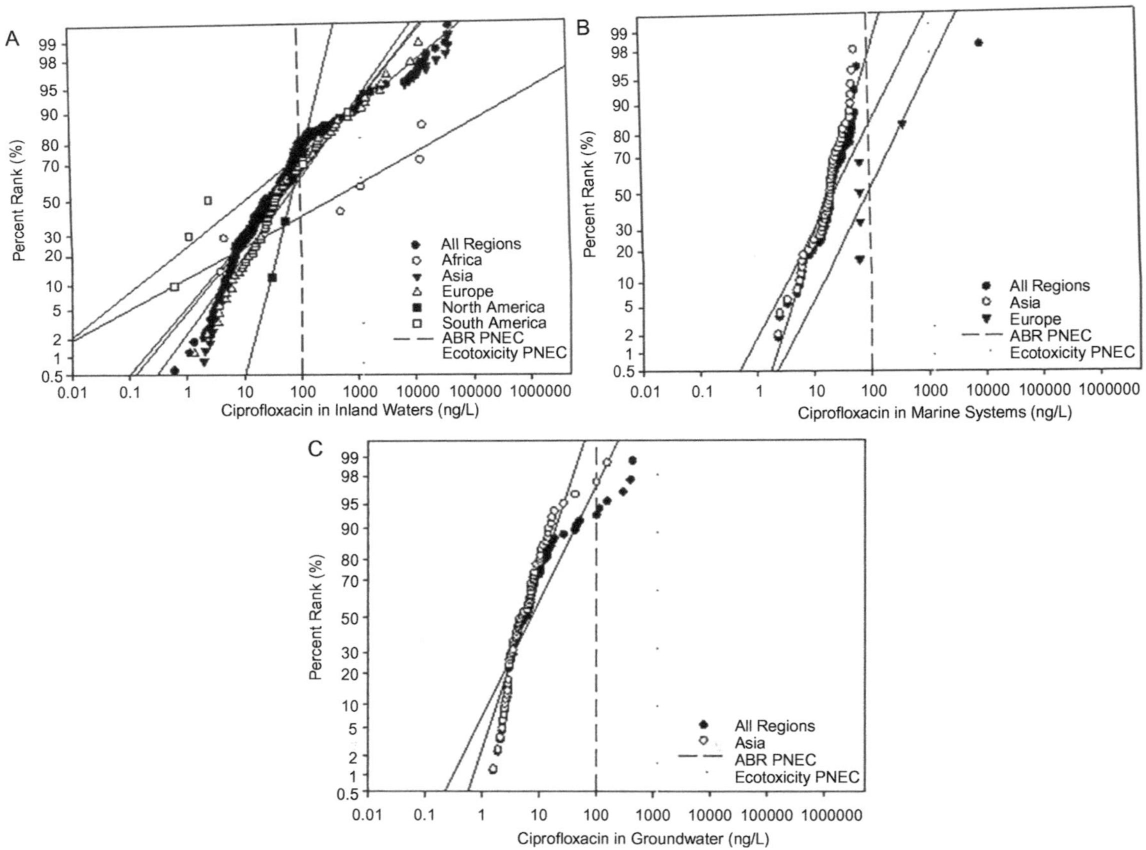

Fig. 3 Measured Environmental Concentration distributions of maximum reported ciprofloxacin in inland waters (A), marine systems (B) and groundwater (C). Vertical lines correspond to Predicted No Effect Concentrations (PNECs) for antibiotic resistance (ABR) development (100 ng/L) and ecotoxicity (1200 ng/L).

population with higher pharmaceutical use, are often located within close range of coastlines.[16] Fifty of the detects were reported in Asia-Pacific, which also has a majority of the world's rapidly growing megacities and increasingly depends on aquaculture. No detects of ciprofloxacin in marine systems were identified from Africa or South America (Tables 1 and 2). However, compared to inland waters, exceedances of only 2% were identified for ABR and ecotoxicity PNECs (Table 3; Fig. 3B).

We further examined ciprofloxacin occurrence in groundwater because surface water contamination can be a major source of contamination to ground water supplies.[34] Almost half of the global population, and 115 million people in the United States alone, rely on groundwater for drinking water.[35] However, this matrix was also less intensively studied than inland waters or effluents, with 111 reported detections of ciprofloxacin (Table 1). These observations were primarily from Asia (96) and Europe[14] with 84 observations from India and China. The maximum concentration of ciprofloxacin in groundwater was 443 ng/L, and an exceedance frequency of the ABR PNEC was identified 7% of the time (Table 3; Fig. 3C). It is particularly important to note that a study in India reported mean concentrations of 14,000 ng/L, 1100 ng/L, and 720 ng/L in groundwater near industrial manufacturing facilities.[36] Risk of antibiotics and ABR development in groundwater, including alluvial aquifers, areas with high surface exchanges, and regions practicing aquifer recharge as a water reuse strategy, represent important areas for future integrated water resources and public health research.

4. CONCLUSIONS

Here we presented a global aquatic hazard assessment of ciprofloxacin in multiple water matrices and identified striking exceedances of previously identified PNECs for ABR and ecotoxicity. For example, exceedances of ABR PNECs were identified 87% and 85% of the time for municipal sewage and effluents, respectively (Table 3; Fig. 2). Exceedances of this ABR threshold was even higher in hospital sewage and effluents (Table 3; Fig. 2). Though such dramatic ABR exceedances were not observed in marine systems or groundwater, these matrices have received much less study and thus deserve additional attention during environmental monitoring and surveillance efforts. It is also important to note that the present study only focused on one common antibiotic; whether such exceedances exist for other anti-infectives employed as veterinary medicines and human pharmaceuticals deserves future study. For example, WWTPs and raw sewage contain complex mixtures of diverse substances, including a variety of antibiotics with diverse mechanisms of action. An understanding of these interactive antibiotic influences on ABR development and adverse health and ecological outcomes should be advanced.

Our findings also highlight the need for coordinated environment and health surveillance in large geographic regions (e.g., South America, Africa) that are markedly understudied, yet are experiencing rapid urbanization and population growth in a number of countries. These continents often have less developed wastewater treatment infrastructure, for which implementation will be a critical component of increased delivery of the essential services of environmental public health.[37] Herein, at a minimum global specimens banks, similar to the U.S. Centers for Disease Controls National Health and Nutritional Examination Survey and the German Environment Agency's Environmental Specimens Bank, should be implemented support environmental monitoring and health surveillance in megacities and particularly in regions strategically advancing water reuse. ABR exceedances of ciprofloxacin identified in the present study, when coupled with several global megatrends, further identify the importance of developing an integrated understanding of urbanization, water, food security and public health in rapidly developing regions.

REFERENCES

1. UNEP. *Global Waste Management Outlook.* United Nations Environment Programme; 2015. https://www.uncclearn.org/sites/default/files/inventory/unep23092015.pdf.
2. Gaw S, Thomas KV, Hutchinson TH. Sources, impacts and trends of pharmaceuticals in the marine and coastal environment. *Philos Trans R Soc Lond B Biol Sci.* 2014;369. 20130572.
3. Kookana RS, Williams M, Boxall ABA, et al. Potential ecological footprints of active pharmaceutical ingredients: an examination of risk factors in low-, middle- and high-income countries. *Philos Trans R Soc Lond B Biol Sci.* 2014;369. 20130586.
4. Furley TH, Brodeur J, Silva de Assis HC, et al. Toward sustainable environmental quality: identifying priority research questions for Latin America. *Integr Environ Assess Manag.* 2018;14:344–357.
5. Boxall AB, Rudd MA, Brooks BW, et al. Pharmaceuticals and personal care products in the environment: what are the big questions? *Environ Health Perspect.* 2012;120:1221–1229.
6. Rudd MA, Ankley GT, Boxall AB, Brooks BW. International scientists priorities for research on pharmaceutical and personal care products in the environment. *Integr Environ Assess Manag.* 2014;10:576–587.
7. Van den Brink PJ, Boxall ABA, Maltby L, et al. Towards sustainable environmental quality: priority research questions for Europe. *Environ Toxicol Chem.* 2018. https://doi.org/10.1002/etc.4205.

8. Brooks BW, Riley TM, Taylor RD. Water quality of effluent-dominated ecosystems: ecotoxicological, hydrological, and management considerations. *Hydrobiologia.* 2006;556:365–379.
9. Rice J, Westerhoff P. High levels of endocrine pollutants in US streams during low flow due to insufficient wastewater dilution. *Nat Geosci.* 2017;10:587–591.
10. Luthy RG, Sedlak DL, Plumlee MH, Austin D, Resh VH. Wastewater-effluent-dominated streams as ecosystem-management tools in a drier climate. *Front Ecol Environ.* 2015;13:477–485.
11. Brooks B, Maul J, Belden J. Antibiotics in aquatic and terrestrial ecosystems. In: Jorgensen SE, Fath B, eds. *Encyclopedia of Ecology*. New York: Elsevier; 2008:210–217.
12. Brain RA, Ramirez AJ, Fulton BA, Chambliss CK, Brooks BW. Herbicidal effects of sulfamethoxazole in lemna gibba: using p-aminobenzoic acid as a biomarker of effect. *Environ Sci Technol.* 2008;42:8965–8970.
13. The Review on Microbial Resistance. *Antimicrobial Resistance: Tackling a Crisis for the Health and Wealth of Nations.* Wellcome Trust and the UK Department of Health; 2014. https://amr-review.org/Publications.html.
14. Pruden A. Balancing water sustainability and public health goals in the face of growing concerns about antibiotic resistance. *Environ Sci Technol.* 2013;48:5–14.
15. Bengtsson-Palme J, Larsson DJ. Concentrations of antibiotics predicted to select for resistant bacteria: proposed limits for environmental regulation. *Environ Int.* 2016;86:140–149.
16. Schafhauser BH, Kristofco LA, Oliveira CMRD, Brooks BW. Global review and analysis of erythromycin in the environment: occurrence, bioaccumulation and antibiotic resistance hazards. *Environ Pollut.* 2018;238:440–451.
17. Larsson DJ, Pedro CD, Paxeus N. Effluent from drug manufactures contains extremely high levels of pharmaceuticals. *J Hazard Mater.* 2007;148:751–755.
18. Kristiansson E, Fick J, Janzon A, et al. Pyrosequencing of antibiotic-contaminated river sediments reveals high levels of resistance and gene transfer elements. *PLoS One.* 2011;6(2), e17038.
19. WHO Advisory Group on Integrated Surveillance of Antimicrobial Resistance. *Critically Important Antimicrobials for Human Medicine: Ranking of Antimicrobial Agents for Risk Management of Antimicrobial Resistance Due to Non-Human Use.* World Health Organization; 2017. http://apps.who.int/iris/bitstream/10665/255027/1/9789241512220-eng.pdf.
20. Redgrave LS, Sutton SB, Webber MA, Piddock LJ. Fluoroquinolone resistance: mechanisms, impact on bacteria, and role in evolutionary success. *Trends Microbiol.* 2014;22:438–445.
21. ACS,. *Ciprofloxacin.* American Chemical Society; 2016. https://www.acs.org/content/acs/en/molecule-of-the-week/archive/c/ciprofloxacin.html.
22. Drlica K. Mechanism of fluoroquinolone action. *Curr Opin Microbiol.* 1999;2:504–508.
23. FDA. *CIPRO (Ciprofloxacin Hydrochloride).* Food and Drug Administration; 2013. https://www.accessdata.fda.gov/drugsatfda_docs/label/2011/019537s074,020780s032lbl.pdf.
24. Kümmerer K, Al-Ahmad A, Mersch-Sundermann V. Biodegradability of some antibiotics, elimination of the genotoxicity and affection of wastewater bacteria in a simple test. *Chemosphere.* 2000;40:701–710.
25. Zou S, Xu W, Zhang R, Tang J, Chen Y, Zhang G. Occurrence and distribution of antibiotics in coastal water of the Bohai Bay, China: impacts of river discharge and aquaculture activities. *Environ Pollut.* 2011;159:2913–2920.
26. Chung SS, Zheng J, Burket SR, Brooks BW. Select antibiotics in leachate from closed and active landfills exceed thresholds for antibiotic resistance development. *Environ Int.* 2018;115:89–96.
27. Corrales J, Kristofco LA, Steele WB, et al. Global assessment of bisphenol A in the environment. *Dose Response.* 2015;13, 155932581559830.

28. Kristofco LA, Brooks BW. Global scanning of antihistamines in the environment: analysis of occurrence and hazards in aquatic systems. *Sci Total Environ.* 2017;592:477–487.
29. Saari GN, Scott WC, Brooks BW. Global scanning assessment of calcium channel blockers in the environment: review and analysis of occurrence, ecotoxicology and hazards in aquatic systems. *Chemosphere.* 2017;189:466–478.
30. Kraupner N, Ebmeyer S, Bengtsson-Palme J, et al. Selective concentration for ciprofloxacin resistance in Escherichia coli grown in complex aquatic bacterial biofilms. *Environ Int.* 2018;116:255–268.
31. Brooks BW, Berninger JP, Ramirez AJ, Huggett DB. Perspectives on human pharmaceuticals in the environment. In: Brooks BW, Huggett DB, eds. *Human Pharmaceuticals in the Environment: Current and Future Perspectives.* New York: Springer; 2012:1–16.
32. Golet EM, Alder AC, Giger W. Environmental exposure and risk assessment of fluoroquinolone antibacterial agents in wastewater and river water of the Glatt Valley watershed, Switzerland. *Environ Sci Technol.* 2002;36:3645–3651.
33. EPA. *Basic Information About Estuaries.* U.S. Environmental Protection Agency; 2016. https://www.epa.gov/nep/basic-information-about-estuaries.
34. Winter TC. Relation of streams, lakes, and wetlands to groundwater flow systems. *Hydrgeol J.* 1999;7:28–45.
35. USGS. *Groundwater Use in the United States.* U.S. Geological Survey; 2017. https://water.usgs.gov/edu/wugw.html.
36. Fick J, Söderström H, Lindberg RH, Phan C, Tysklind M, Larsson DGJ. Contamination of surface, ground, and drinking water from pharmaceutical production. *Environ Toxicol Chem.* 2009;28:2522.
37. CDC. *10 Essential Environmental Public Health Services.* Atlanta, GA, USA: Centers for Disease Control and Prevention; 2018. http://www.cdc.gov/nceh/ehs/10-essential-services/index.html. Accessed 4 July 2018.

CHAPTER FOUR

Group I Intron-Based Therapeutics Through *Trans*-Splicing Reaction

Chang Ho Lee*, Seung Ryul Han†, Seong-Wook Lee*,†,1

*Department of Integrated Life Sciences, Dankook University, Yongin, Republic of Korea

†Rznomics Inc., Gwangju, Republic of Korea

[1]Corresponding author: e-mail address: SWL0208@dankook.ac.kr

Contents

1. Introduction 80
2. Group I Intron 82
   2.1 Self-Splicing Activity of Group I Intron 82
   2.2 Development of *Trans*-Splicing Group I Ribozyme 84
3. Group I Intron as Therapeutics 86
   3.1 *Trans*-Splicing Ribozyme for RNA Repair 86
   3.2 *Trans*-Splicing Ribozyme for RNA Reprogramming 87
4. Concluding Remarks 93
Acknowledgments 94
References 94

Abstract

In 1982, the Cech group discovered that an intron structure in an rRNA precursor of *Tetrahymena thermophila* is sufficient to complete splicing without assistance from proteins. This was the first moment that scientists recognized RNAs can have catalytic activities derived from their own unique three-dimensional structures and thus play more various roles in biological processes than thought before. Several additional catalytic RNAs, called ribozymes, were subsequently identified in nature followed by intense studies to reveal their mechanisms of action and to engineer them for use in fields such as molecular cell biology, therapeutics, imaging, *etc.* Naturally occurring RNA-targeting ribozymes can be broadly classified into two categories by their abilities: *Self*-cleavage and *self*-splicing. Since ribozymes use base-pairing to recognize cleavage sites, identification of the catalytic center of naturally occurring ribozymes enables to engineer from "*self*" to "*trans*" acting ones which has accelerated to design and use ribozyme as valuable tools in gene therapy fields. Especially, group I intron-based *trans*-splicing ribozyme has unique property to use as a gene therapeutic agent. It can destroy and simultaneously repair (and/or reprogram) target RNAs to yield the desired therapeutic RNAs, maintaining endogenous spatial and temporal gene regulation of target RNAs. There have been progressive improvements in *trans*-splicing

Progress in Molecular Biology and Translational Science, Volume 159
ISSN 1877-1173
https://doi.org/10.1016/bs.pmbts.2018.07.001

ribozymes and successful applications of these elements in gene therapy and molecular imaging approaches for various pathogenic conditions. In this chapter, current status of *trans*-splicing ribozyme therapeutics, focusing on *Tetrahymena* group I intron-based ribozymes, and their future prospects will be discussed.

1. INTRODUCTION

Gene therapy includes the delivery of therapeutic nucleic acids (DNA or RNA) as a drug for the treatment, cure, or prevention of various human diseases. Gene therapy has represented an attractive strategy to the substitution or suppression of defective genes responsible for disease development through carrying wild-type or therapeutic genes such as siRNA, antisense oligonucleotide (ASO), or ribozymes to cells lacking the endogenous genes or harboring malfunctional copies of genes, respectively.[1–4] However, such traditional gene therapeutic approaches would likely fail in a dominant mutation-related disease because it would be difficult to completely eliminate the function of mutated allele through simple delivery of wild-type copies of genes. As an alternative, it is recommended to use siRNAs or ASOs, which can target and decrease the expression of pathogenic mutated genes, for the treatment of the dominant mutation-related diseases.[5, 6] However, its use as a gene therapy tool has been limited because of some drawbacks, such as the requirement of a delivery vehicle, off-target effects, potential immunogenicity, and toxicity.[7, 8]

From this perspective, the most ideal approach for gene therapy against several types of diseases may be a gene replacement or alteration which can precisely maintain spatial and temporal regulation of gene expression in cells and tissues. To achieve this, a direct genome targeting approach has been developed to edit disease-causing genes to normal ones with zinc finger nucleases (ZFN) and transcription activator-like effector nucleases (TALEN) through inducing double-strand breaks (DSB) at the corresponding DNA target site. The resulting DSBs can be fixed either by non-homologous end joining (NHEJ) or homology-directed repair (HDR) pathways. NHEJ pathway produces insertions and deletions (indels) to induce a permanent gene knockout. In contrast, HDR pathway utilizes a DNA template homologous to regions neighboring the DSB site for either amendment of an existing gene or complement of a new gene to the targeted genome.[9] Recently, a precise and programmable RNA-guided DNA endonuclease system, called clustered regularly interspaced short palindromic repeats (CRISPR) and

CRISPR-associated (Cas) genes (CRISPR/Cas) system derived from a bacterial adoptive immune network, has been successfully developed for site-specific genome engineering.[10–12] Compared to ZFN and TALEN, the CRISPR/Cas9 system can be more easily engineered and more efficiently induces DSBs and thus offers more genome editing opportunities.[9, 11, 13–15] The CRISPR/Cas system consists of two components forming a single guide RNA (sgRNA)-Cas9 endonuclease complex.[13] The sgRNA leads the Cas9 to target DNA site and induces DSB followed by a 3- to 5-nucleotide protospacer-adjacent motif (PAM).[16] Since it is only required to change the sgRNA according to the target, this system can be simply and cost-effectively engineered to target the wanted genomic loci and can simultaneously target multiple loci in the genome with multiple sgRNAs.[17–19] However, there are still limitations to overcome for its successful application to gene therapy, such as delivery methods,[15, 20, 21] off-target effects,[22] potential immunogenicity, and low genome editing efficiency particularly when using HDR pathway.[23] Even more, gene editing with programmable nucleases, including the CRISPR/Cas system, may not be effective against diseases resulting from an alteration in the RNA pattern due to alternative splicing, defective RNA modification, and overexpression. Therefore, for targeting and replacing and/or reprograming disease-causing RNAs, specific, efficient, and safe tool is needed to be developed.

RNA-targeting strategies with siRNA, ASOs, and ribozyme use Watson–Crick base-pairing to recognize target RNAs and inhibit translation. However, each approach removes target RNAs *via* its own mechanism.[6] ASOs and siRNA depend on cellular proteins to exert their activities. On the other hand, ribozymes fold into distinct tertiary structures which give catalytic abilities without assistance from proteins.[5, 24] Since the initial discovery in the early 1980s,[25–27] numerous ribozymes have been discovered and artificial ones developed as valuable tools in the fields of molecular and cell biology and therapeutics. Ribozymes are characterized as cleaving or splicing types.[28–30] Both types have been modified and developed to recognize RNAs in *trans*. ASOs, siRNAs, and *trans*-cleaving ribozymes contain activity of only removing target transcripts in cells. In contrast, *trans*-splicing ribozymes enable RNA editing process through eliminating and simultaneously repairing or reprograming target RNA without modification of either the sequence and expression regulation of the corresponding gene.[5, 6, 20, 31] Furthermore, in contrast to ASO- and siRNA-based approaches and genome editing approaches, *trans*-splicing ribozyme would not necessitate any kinds of endogenous and/or exogenous proteins that could induce cellular toxicity and potential immunogenicity. In addition, control of target binding length or post-transcriptional regulation could enhance the ribozyme specificity and thus minimize off-target effects. RNA repair and/or reprogramming ability of *trans*-splicing ribozymes have facilitated the ribozymes as tools of gene therapy[32–44] and molecular imaging[45–48] against diverse human disease. In the following sections, we will describe the current strategies and status of gene therapy applications of *trans*-splicing ribozymes, mainly focused on the *Tetrahymena* group I intron-derived types.

2. GROUP I INTRON

In 1982, the group I intron was discovered from an rRNA transcript of *Tetrahymena thermophila* as the first ribozyme, which can self-splice from a primary transcript without the help of any proteins.[27] Since its discovery, many other self-splicing group I introns have been identified in tRNAs, mRNAs, and rRNAs of numerous organisms and viral genomes.[49, 50] The three-dimensional structures of catalytic cores of three different group I introns are strikingly similar in spite of poor conservation of their primary sequences.[51–56] The catalytic core of the group I intron is present at the junction of the P4–P6 (P4, P5, P6) and P3–P9 (P3, P7, P8, P9) domains. P1 and P10 define the 5′ and 3′ splice sites, respectively, which are formed by base-pairing between an internal guide sequence (IGS) located immediately downstream of the 5′ splice site and an exon sequence flanking the splice site. P1 helix is important for determining the 5′ splice site because it is defined by G·U wobble base-pairing followed by the base-pairing of 4–6 nucleotides between the IGS of intron and the 3′ end of the 5′-exon in the P1 helix.[49, 57, 58] The last nucleotide of the intron, ωG, and the first nucleotide of the 3′-exon define the 3′ splice site, and there are interactions between the adjacent portion of IGS and the first nucleotide of the 3′-exon in the P10 helix.[49, 52, 59, 60] Other components surrounding the conserved catalytic core are important for ribozyme stability and correct folding (Fig. 1A).[61–64]

2.1 Self-Splicing Activity of Group I Intron

Self-splicing reaction of group I intron is a two-step transesterification reaction, resulting in the removal of the intronic sequences and ligation of the flanking exons (Fig. 1B).[65] During these process, group I intron only requires Mg^{2+} and guanosine without any assistance of proteins.[66, 67]

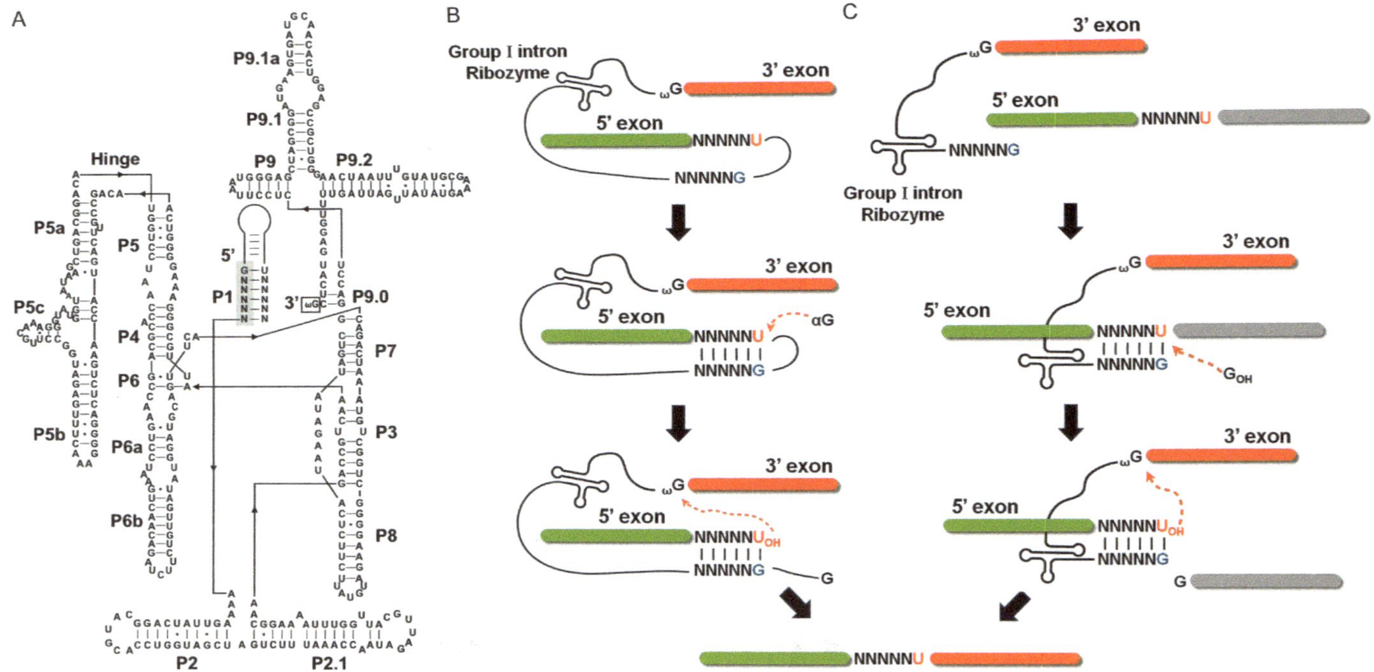

Fig. 1 *Tetrahymena thermophila* group I intron. (A) Secondary structure and sequence of the group I intron. The *gray box* indicates the internal guide sequence (IGS). The *open box* indicates the omega-G (ωG). Mechanism of (B) self-splicing and (C) *trans*-splicing reaction of the group I intron.

The reaction is initiated when the P1 helix correctly folds into the catalytic core of the intron in response to the binding of exogenous guanosine to a site within the folded catalytic core.[49, 52, 68, 69] The 3′ hydroxyl group (3′-OH) of the bound exogenous guanosine acts as a nucleophile and attacks phosphate at the 5′ splice site resulting in the covalent linkage of exogenous guanosine to the 5′ end of the intron and the generation of a free 3′-OH on the uridine at the 3′ terminus of the 5′-exon. Subsequently, ωG at the 3′ end of the intron substitutes exogenous guanosine through a conformational change assisted by P10 helix formation.[54] In this way, the 3′ splice site is positioned for nucleophilic attack by the 3′-OH of the 5′-exon, resulting in the ligation of the adjoining exons and the release of the intron.[27]

2.2 Development of *Trans*-Splicing Group I Ribozyme

Group I ribozyme from *T. thermophila* can be modified to perform the splicing reaction in *trans*, resulting in the ligation of the 5′-exon of one RNA with the 3′-exon of another RNA *via* similar mechanism to self-splicing reaction (Fig. 1C).[57, 70, 71] *Trans*-splicing group I ribozyme recognizes target site on the substrate RNA through base-pairing with its IGS (Internal Guide Sequence), shaping P1 helix. The only sequence in the target RNA required for binding is a uridine residue just upstream the 5′ splice site that forms a conserved G·U base pair with a guanosine in the IGS.[57] Since the only sequence requirement for its activity is a uridine preceding the 5′ splice site in the target RNA, self-splicing ribozymes can be engineered to *trans*-splicing ones for repair or reprogram of disease-related genes to corresponding normal, or entirely new genes for therapeutic purposes.[29, 31, 72–74] The Sullenger group first described that *trans*-splicing reactions can occur in *Escherichia coli*[71] and mammalian cells[75] using *T. thermophila*-derived ribozyme, resulting in the repair of mutant *lacZ* RNA.

Although the uridine residue is the only sequence required for targeting of *trans*-splicing ribozymes, all the uridines within the substrate RNA may not be accessible,[76, 77] because RNAs exist in multiple distinct conformations and/or interact with various proteins in cells. Therefore, the identification of accessible uridine sites in target RNAs is necessary for developing an efficient *trans*-splicing ribozyme.[76, 77] To identify efficient splice sites on substrate RNAs in cells as well as *in vitro*, an RNA mapping strategy was developed using a ribozyme library with randomized IGS (5′-GNNNNN-3′) and 3′-exon tag sequence.[75, 78] The ribozyme library reacts with the substrate

RNAs, and accessible uridines can be identified by RT-PCR and subsequent sequencing of the *trans*-splicing chimeric products. Recent advances in computational biology allow prediction of accessible uridines in substrate RNAs by calculating the binding free energies between target sites and ribozyme IGS. This computational prediction method can avoid artifacts such as the preferred transcription of library molecules with purines at the 5′-end[79] and biases from the preferential PCR amplification of shorter amplicons.[80] In this respect, computational prediction may have advantages over the experimental RNA mapping strategy,[81] although prediction of accessible sites in target RNA complexed with proteins is difficult.

The group I ribozymes with only a 6-nt-long IGS were relatively inactive and non-specifically targeting when expressed in mammalian cells.[33, 75] Therefore, the ribozyme needs to be optimized for the therapeutic application. Wild-type exon sequences flanking the natural group I intron were shown to promote accurate folding and stability of active ribozyme, indicating that the sequence peripheral to the ribozyme and/or corresponding RNA structure may be needed to be modified for optimization.[82–84] Accordingly, extended base-pairing between IGS and target RNA and promoting P10 helix formation of *trans*-splicing ribozyme have been proved to increase the efficiency and specificity of the *trans*-splicing ribozyme.[66, 70, 85, 86] Improvement of specificity of intracellular *trans*-splicing reaction and no off-target effects by the modified ribozymes were verified using 5′-end rapid amplification of cDNA ends (RACE) PCR analysis of the total *trans*-spliced products in the target RNA-expressing cells.[37, 38, 87, 88] Extended base-pairing comprises P1 extension of about three base-pairs, an internal loop, and a further duplex between a part of the extended guide sequence (EGS), termed antisense region, and the substrate RNA.[70, 89, 90] The optimal length of P1 extension is exactly three base-pairs which corresponds to the length of the natural P1 extension in the *T. thermophila* group I intron.[57] The optimal length of EGS for enhancement of the efficacy and specificity of a *trans*-splicing ribozyme depends on the target RNA and thus requires experimental determination.[33, 70, 90, 91] Although the optimal length of the P10 helix appears to differ between splice sites, in most cases, an approximate six base-pair in the P10 helix can efficiently increase the 3′-splice site specificity.[70, 85] Recently, to develop improved ribozyme variants without individually constructing and testing diverse ribozyme derivatives, selection[86, 90, 92–96] and evolution[97–99] methods have been used to identify and develop *trans*-splicing ribozymes with improved activity.[100]

3. GROUP I INTRON AS THERAPEUTICS

The *trans*-splicing ribozyme can edit target RNAs through elimination and simultaneous replacement of disease-related RNA with any desired RNA. Consequently, this technique has been explored in the milieu of therapeutic and molecular imaging fields. There are two types of applications using *trans*-splicing ribozyme: (1) repair of a mutation in the 3′-portion of the substrate RNA through replacement of the substrate 3′-exon with the ribozyme 3′-exon encoding the wild-type sequence and (2) reprogramming through replacement of the substrate 3′-exon with a new, unrelated ribozyme 3′-exon encoding a therapeutic or a reporter sequence. Although group I intron ribozymes derived from *Pneumocystis carinii*,[101] *Candida albicans*,[102] and the *myxomycetes Didymium* and *Fuligo*[103] have been engineered to perform *trans*-splicing reactions, most *trans*-splicing ribozymes developed for therapeutic purposes are based on the *T. thermophila*-derived group I intron.

3.1 *Trans*-Splicing Ribozyme for RNA Repair

Trans-splicing ribozymes for RNA repair have been designed to target and cleave upstream of mutated genes and subsequently replace them with ribozyme 3′-exon RNA encoding wild-type sequence to restore the correct gene function.[71] Any uridine residue in a mutated RNA could be recognized by a *trans*-splicing ribozyme, indicating that most of RNA-level genetic mutations found in genetic disease, cancer, or others can be targeted and repaired to produce properly functioning genes without disturbing endogenous spatial and temporal gene regulation.[33, 36, 39, 40, 42, 104, 105]

Two studies were reported for the first time in 1998 that *trans*-splicing ribozymes could be engineered to repair clinically relevant mutant transcripts associated with genetic diseases. Repair of a trinucleotide repeat in myotonic dystrophy protein kinase (DMPK) associated with myotonic dystrophy was reported by Phylactou et al., who developed a *trans*-splicing ribozyme to bind immediately upstream of the mutated CUG repeat expansion area of DMPK mRNA and replace the 3′ end with the wild-type RNA sequence *in vitro* and in human fibroblasts.[39] Repair of sickle beta-globin (β^S-globin) mRNA into transcripts with normal functions was accomplished by the Sullenger group, which designed a Rib61–3′γ *trans*-splicing ribozyme to target and replace the mutant β^S-globin mRNA responsible for sickle cell anemia with an mRNA encoding anti-sickling γ-globin *in vitro* and in erythrocyte precursors derived from sickle cell patients.[104] It has been known that a 10–20% level of γ-globin expression relative to β^S-globin RNAs in a patient's red blood cells can greatly improve the clinical prognosis.[106, 107] In a following study, for the achievement of clinically amenable repair efficiency, the Sullenger group optimized *trans*-splicing ribozyme expression cassettes and enhanced repair rates up to 10–50% of sickle β^S-globin RNAs in transfected mammalian cells.[33] However, the study used a cytoplasmic T7 RNA polymerase expression cassette for high expression of the ribozyme, which cannot be used in therapeutic applications. Optimization of vector systems to reach sufficient ribozyme expression levels or highly efficient ribozyme derivatives will be required for therapeutic application of the ribozyme. Rogers et al. developed a *trans*-splicing ribozyme which can repair a mutant canine skeletal muscle chloride channel (cClC-1) mRNA causing the inherited disorder myotonia congenita.[40] Although a repair efficiency of this ribozyme with approximately 1.2%, 18% of cells were repaired exhibiting significant restoration of chloride channel function. Group I introns derived from *Didymium* and *Fuligo* exhibited equivalent *trans*-splicing reaction efficiency as those of the *Tetrahymena* and facilitated the repair of mutated α-mannosidase RNA *in vitro*.[103]

Trans-splicing ribozymes have been designed to repair mutant genes associated with cancers.[36, 42, 105] Watanabe et al. developed *trans*-splicing ribozymes targeting uridine at positions 41 and 65 of a mutant p53 tumor suppressor gene, thus repairing the mutant p53 RNA and producing a functional p53 protein with restored promoter activity in various p53 mutated cancer cell lines.[105] Our group also developed a *trans*-splicing ribozyme targeting uridine at position 96 of mutant p53 RNA with a repair efficiency of approximately 10%, consequently inducing p53-responsive genes p21 and Bax and efficiently triggering apoptosis in transfected ovarian cancer cell lines.[42] Kastanos et al. described a *trans*-splicing ribozyme which can repair a two base-deleted p16 transcript in a pancreatic cancer cell line, restoring normal p16 protein synthesis and consequently reducing growth rate of transfected cells.[36]

3.2 *Trans*-Splicing Ribozyme for RNA Reprogramming

Trans-splicing ribozyme-mediated RNA repair is a valuable and hopeful approach for treating genetic and some malignant diseases. However, it may be not applicable to other diseases. For instance, in a virus infection-related disease, viral transcripts must be removed and replaced with a therapeutic gene to ensure the clearance or cure of virus-infected cells.

Moreover, multiple pathways are dysregulated in most cancers, indicating that multiple genes such as oncogenes and tumor suppressor genes are mutated combined with alteration of expression levels and thus single gene repair would not be effective. To overcome this limitation, our group and others engineered *trans*-splicing ribozymes to reprogram viral transcripts[35, 41, 77, 108] or tumor-related genes[32, 37, 38, 44, 47, 87, 109–113] to induce cell death and clear virus-infected or cancer cells. The ribozyme-mediated reprogramming triggers therapeutic gene expression only in virus-infected or cancer cells as long as that the target RNA is expressed only in the cells, exhibiting no off-target effects such as targeting and splicing with unrelated RNAs. Of note, the ribozyme simultaneously reduces the expression of the target gene, which could provide an additive therapeutic effect. Furthermore, *trans*-splicing ribozymes have been explored as molecular imaging tools by visualizing the target RNA-expressing cells through reprogram of target RNA with 3′-exons with reporter gene activity or enzymatic activity using positron emission tomography-computed tomography (PET-CT). This approach could be used for the detection of specific RNA expression and simultaneous image-guided therapy in living organisms.[45, 46, 48]

Our group developed *trans*-splicing ribozymes to reprogram hepatitis C virus (HCV) and selectively kill the virus-infected cells.[41] We chose the 5′ untranslated region (5′-UTR) of HCV as the target, as this site harbors an internal ribosome entry site (IRES) that is highly conserved among viral genotypes[114] and critical for the cap-independent viral translation for HCV replication.[115] We identified the most accessible uridine site (position 199) in HCV IRES RNA using *in vitro* RNA mapping method, constructed *trans*-splicing ribozyme recognizing the site, and verified the accessibility of this site in cells by the ribozyme.[77] The efficient replacement of the target RNA with therapeutic RNA encoding diphtheria toxin A chain (DTA) was observed, followed by induction of the selective killing of HCV RNA-containing cells.[41] Carter et al. designed an anti-dengue virus (DENV) *trans*-splicing ribozyme to target the uridine at position 143 in the highly conserved 5′–3′ cyclization sequence (CS) region common to all DENV serotypes.[108] The ribozyme was designed to replace the target RNA with therapeutic RNA encoding deletion of the Bax BH-3 domain (ΔN Bax) which can more rapidly initiate irreversible apoptosis than Bax.[116] This ribozyme reprogrammed the CS regions of all DENV serotypes, and subsequently induce apoptosis during DENV infection.[34] RNA reprogramming approach to the viral-infected cells could be useful because the ribozyme can target and reduce the viral RNAs and concurrently and selectively transfer therapeutic genes to treat or kill virus-infected cells without causing side-effects in healthy cells. Theoretically any RNAs can be targeted by *trans*-splicing ribozymes, and thus in addition to viral RNAs of positive-sensed RNA viruses which were targeted so far, *trans*-splicing ribozymes could be developed to recognize and target other RNA viral genomes and transcripts of DNA viruses also for control of the virus infection.

Trans-splicing ribozymes to repair defective genes can be effective for cancers caused by one or a few mutations. However, as the number of mutations within several key genes, including proto-oncogenes, tumor suppressor genes, and DNA repair genes, increase in most of cancers, the ability to simultaneously treat all of them decreases. Therefore, more effective and reliable methods are required to treat polygenic disease such as most of cancers. So far, the most extensively developed therapeutic *trans*-splicing ribozymes against cancer are ones that can reprogram highly expressed RNAs in the cancer cells into RNAs encoding suicide gene that triggers cancer cell death. This reprogramming approach might provide an effective cancer gene therapy tool because the expression of a cancer-specific suicide gene targeting a single RNA can efficiently induce death of cancer cells regardless of the involvement of numerous mutations in the cells.

Our group developed a *trans*-splicing ribozyme targeting human telomerase reverse transcriptase (hTERT) RNA. The ribozyme was designed to reprogram the target RNAs into RNAs encoding herpes simplex virus thymidine kinase (HSV-tk) which activity converts the prodrug ganciclovir (GCV) to a cytotoxin specifically in cancer cells expressing the target RNA.[35, 38, 43, 112, 117–119] Telomerase activity has been found in approximately 80–90% of human cancer which is regulated mainly by hTERT mRNA[120–122] and most normal human somatic tissues express much lower or undetectable levels of telomerase activity,[123] indicating that hTERT is an attractive therapeutic target in most of cancers.[124] We used an *in vitro* mapping strategy to identify the most accessible uridine site (position 21) within the hTERT RNA sequence and constructed a *trans*-splicing ribozyme containing the HSV-tk gene as a 3′-exon (Ribo-HSVtk).[38] Intracellular expression of Ribo-HSVtk specifically and efficiently induced hTERT RNA reprograming and triggered cytotoxic activity only in hTERT-expressing cancer cells treated with GCV. Adenoviral hTERT-targeting *trans*-splicing ribozyme with a downstream reporter gene (Ad-Ribo-LacZ) or suicide gene (Ad-Ribo-HSVtk) was generated and used for delivery into human cancer cell lines and human colon cancer xenograft models.[110, 111] Ad-Ribo-LacZ and Ad-Ribo-HSVtk reduced hTERT mRNA levels and selectively induced β-galactosidase expression and regressed tumor progression in response to GCV, respectively, in a hTERT-dependent manner in all tested hTERT-positive cancer cell lines and xenograft models. However, the expression of the hTERT-targeting *trans*-splicing ribozyme needs to be restricted to specific tissues because of plausible expression of hTERT mRNA in highly proliferating normal cells. Our group attempted to induce anticancer gene activity specifically in liver cancer cells through tissue-specific expression of the hTERT-targeting *trans*-splicing ribozyme.[44] We tested liver-specific promoters, such as promoters of apolipoprotein E (ApoE), phosphoenolpyruvate carboxykinase (PEPCK), or albumin, to restrict the induction of *trans*-splicing ribozyme activity to liver cancer cells expressing hTERT mRNA. In subsequent study, we verified that the systemic delivery of adenoviral hTERT-targeting *trans*-splicing ribozyme under the control of the liver-specific PEPCK promoter (Ad-PEPCK.Ribo-LacZ) into a hepatocellular carcinoma (HCC)-derived peritoneal carcinomatosis model reduced the hTERT RNA level up to 25% compared to control and induced changes in global gene expression profiles associated with apoptosis, angiogenesis, and metastasis.[117] Of note, we observed that Ad-PEPCK.Ribo-HSVTK efficiently decreased hTERT RNA levels (~75%) and caused hTERT-positive liver tumor regressions with least liver toxicity in the HCC-derived model of peritoneal carcinomatosis.[112] Furthermore, Kim et al. reported that the delivery of more improved Ad-PEPCK.Ribo-HSVTK generated efficient antitumor efficacy and an improved specificity of HCC targeting without damaging normal hepatocytes in a mouse xenograft model of orthotopic and multifocal HCC.[46] Moreover, Kim et al. found cell damage responses as well as instability of telomere integrity by Ad-Ribo-HSVTK-mediated hTERT downregulation, indicating that decreased level of hTERT in addition to selective expression HSV-TK amplified the anticancer efficacy of the hTERT-targeting *trans*-splicing ribozyme.[46] These results indicate the usefulness of the reprograming approach with a hTERT-targeting *trans*-splicing ribozyme for anticancer therapy. Phase I clinical trial is currently being employed with Ad-Ribo-HSVtk for the Korean patients with hepatic metastases of solid tumors.[46]

Since hTERT level is increased in hematopoietic progenitor cells, germline cells, and the renewal tissues including the epidermis, intestinal crypts, and regenerating hepatocytes,[124] suicide gene therapy with hTERT-targeting *trans*-splicing ribozymes could cause harmfulness also to such normal stem and regenerating cells. Our group inserted complementary binding sites for microRNA-181a (miR-181a), which is specifically expressed in hematopoietic cells,[125–127] in the 3′ untranslated region (3′-UTR) of the hTERT-targeting *trans*-splicing ribozyme to prevent its potential side effects to hTERT(+) hematopoietic stem cell-derived blood cells.[128] This ribozyme selectively induced cytotoxicity in hTERT-positive but miR-181a-negative cancer cell lines, resulting in the regression of hTERT(+) liver tumors with high efficacy and minimal liver toxicity in a mouse xenograft model of orthotopic and multifocal HCC.[128] Our group further designed hTERT-targeting *trans*-splicing ribozyme with hepatocyte-specific microRNA-122a (miR-122a) binding sites at its 3′-UTR to regulate activity of the ribozyme in a microRNA-dependent manner, and thus to improve HCC-specificity of the ribozyme with least side effects in the regenerating non-tumor liver tissues of HCC patients.[129] The hepatocyte-specific miR-122a was selected because it is known to be highly down-regulated particularly in poorly differentiated and aggressive HCC.[130] Adenoviral delivery of the ribozyme induced an anticancer effect with high efficacy and minimal hepatotoxicity while controlling ribozyme expression in a miR-122a-dependent manner in xenografted orthotopic murine models of multifocal HCC.[129] Our group also constructed mTERT-Ribo-TK, a mouse TERT-targeting *trans*-splicing ribozyme containing the HSV-tk gene as a 3′-exon, to test the anticancer effects of the ribozyme in a syngeneic immunocompetent HCC mouse model.[118] Adenoviral delivery of the ribozyme with miR-122a binding sites specifically retarded the survival of mTERT-expressing HCC cells and reduced the tumor burden with high efficacy and regulated ribozyme expression without non-specific cytotoxicity in normal liver tissues in the syngeneic orthotopic murine models of multifocal HCC model.[129] In addition to the endogenous microRNA control of *trans*-splicing ribozyme activity, our group developed an exogenously and allosterically controllable hTERT-targeting *trans*-splicing ribozyme.[131] The *trans*-splicing ribozyme activity could be controlled by exogenously treated theophylline *in vitro* and in cells as well when the terminal loop regions of both P6 and P8 domains of the hTERT-targeting *trans*-splicing ribozyme were substituted with theophylline-specific RNA aptamers as sensor domains. This result indicates that aptamer-appended *trans*-splicing ribozyme could be designed to be spatially and temporally controlled with endogenous or extracellularly-treated small molecules.

Other RNAs overexpressed in cancer cells, such as carcinoembryonic antigen (CEA),[87] cytoskeleton-associated protein 2 (CKAP2),[32, 88] alpha-fetoprotein (AFP),[109] and pancreatic adenocarcinoma up-regulated factor (PAUF) RNA,[47] were also targeted for the development of anticancer *trans*-splicing ribozymes which can induce therapeutic transgene expression specifically in the target RNA-expressing cancer cells through RNA reprograming. Tumor-specific splicing variants and mutations in cancer cells were also targeted for *trans*-splicing ribozyme-mediated reprograming and subsequent induction of anticancer activity.[37, 44] For example, our group developed a *trans*-splicing ribozyme specific for a splicing RNA variant of AIMP2 lacking exon 2 (AIMP2-DX2) which are highly expressed in human lung cancer cells and tissues of cancer patients.[132] The ribozyme was observed to selectively stimulate HSV-tk activity in AIMP2-DX2 RNA-expressing lung cancer cells through RNA reprogramming, resulting in the retardation of cancer cell growth with high specificity and efficiency in response to GCV treatment.[44] We also developed a *trans*-splicing ribozyme to specifically target and selectively reprogram RNA encoding the KRAS G12V mutation,[37] which causes constitutive RAS activation in many kinds of human cancers.[133] Adenoviral delivery of the ribozyme specifically downregulated KRAS G12 V expression, selectively induced to regression of KRAS G12 V-harboring cancer cells, and retarded the growth of xenografted tumors harboring the KRAS G12 V mutation but had no effects on cells and xenografted tumors harboring wild-type KRAS. This study indicates that *trans*-splicing ribozymes can be designed to repair or reprogram target RNAs with high specificity and accuracy, discriminating targets even at the level of a single nucleotide difference.

Of note, we observed that mTERT-specific *trans*-splicing ribozyme induced tumor cell lysis and elicited local and systemic antitumor immunity,[134, 135] and completely inhibited a secondary tumor challenge in a syngeneic murine model of orthotopic HCC.[129] Shin et al. constructed a recombinant mTERT-Ribo-TK vector integrated with cDNA encoding a recombinant fusion protein comprising soluble programmed death-1 (sPD1) and the Fc portion of mouse IgG2a (sPD1-Ig) for the purpose of blocking PD-L1 in the tumor microenvironment.[136] Adenoviral delivery of the TERT-targeting ribozyme, then, induced tumor cell lysis, followed by sPD1 Ig-mediated priming of CD8+ T cells specific for the tumor antigen *via* dendritic cells (DC). The ribozyme more efficiently elicited antitumor activity and completely suppressed a secondary tumor challenge at a distal site.[136] These studies indicate that efficient antitumor strategy could be achieved through immune enhancement combined with RNA reprograming.

Target RNA-specific reprograming approach with *trans*-splicing ribozymes could be also applied to the molecular imaging of target RNA expression in live animals.[45–48] Hong et al. developed a *trans*-splicing ribozyme to monitor mouse CKAP2 (mCKAP2) RNA expression in tumors in murine orthotopic HCC model *via* reprogramming the target RNA with RNA encoding a luciferase reporter gene.[45] *Trans*-splicing ribozyme with reporter gene was also developed to monitor tumor-specific mutant p53 mRNA in mice and to visualize *in vivo* image of siRNA-mediated gene silencing.[48] Moreover, Ad-PEPCK.Ribo-HSVTK and tracer 9-(4-[18F]fluoro-3-hydroxymethylbutyl) guanine ([18F]FHBG) were successfully used in combination for PET to estimate delivery efficacy, detect transgene expression at target sites, and thus, monitor the anticancer effectiveness of the *trans*-splicing ribozyme *in vivo*.[46] The capability of non-invasive imaging of disease-related mRNAs and simultaneous monitoring of therapeutic RNA expression and therapeutic effects would be critical to apply *trans*-splicing ribozyme in patients.

4. CONCLUDING REMARKS

RNA repair and reprogramming approaches with group I intron-based *trans*-splicing ribozyme can reduce disease-associated target RNA levels and simultaneously induce therapeutic gene activity specifically in cells expressing the target RNA. In addition, the spatial and temporal regulation of endogenous transcripts can be maintained within a cell treated with *trans*-splicing ribozymes. Moreover, intracellular *trans*-splicing reaction of the ribozyme would not require the help of endogenous and/or exogenous proteins that could induce toxicity and immunogenicity. These indicate that the ribozyme-mediated approaches would be more therapeutically beneficial than many traditional therapeutic modalities. Accordingly, *trans*-splicing ribozymes have been developed against various genetic, infectious, and malignant diseases and their efficacy and specificity have been tested in different cell types. Particularly, the therapeutic efficacy of *trans*-splicing ribozymes targeting tumor-related or causing RNAs were successfully verified in diverse animal tumor models. Accordingly, clinical trials have recently been initiated in Korea to test the hTERT-targeting *trans*-splicing ribozyme in cancer patients.[46]

Various methods have been attempted to modify and optimize *trans*-splicing ribozymes for therapeutic purposes. Many studies proved that modifications of *T. thermophila* group I intron ribozyme such as IGS extension and P10 promotion enhances the efficacy and specificity of the ribozyme with minimal off-target effects in cells.[71, 76, 85, 91] Moreover, the *in vivo* efficacy of the *trans*-splicing ribozyme was shown to depend on the expression levels of the target RNA and ribozyme.[119] Localized and efficient delivery to target RNA-expressing cells using a recombinant adenoviral vector and tissue-specific promoter system generated selective and efficient expression of *trans*-splicing ribozymes in target cells and target tissues to successfully promote target RNA reprograming, thus exerting the desired therapeutic effects in various cancer models.[38, 43, 111, 112] *Trans*-splicing ribozymes were further modified in order for their activities to be more precisely and specifically regulated by endogenous tissue-specific miRNAs[128, 129] or exogenously inoculated small molecules.[131]

To generally apply *trans*-splicing ribozyme in clinical circumstances, the following issues should be considered. (1) Efficient delivery/vector systems are needed for its therapeutic efficacy, as for other gene therapy approaches. The expression level of *trans*-splicing ribozyme is directly proportional to its intracellular activity,[119] indicating that targeted and efficient delivery and expression of the ribozyme is critical for its therapeutic capability. (2) Repeated treatment would be needed in some cases of diseases for the achievement of therapeutic effects. RNA targeting with *trans*-splicing ribozyme only transiently produces reprogrammed therapeutic gene activity. Therefore, the continual delivery or expression of the ribozyme in target cells would probably be desirable in some diseases such as genetic diseases. (3) Limitations may exist in the choice of sites in target RNA recognized by *trans*-splicing ribozymes with efficiency sufficient to reach therapeutic potentials due to target RNA structure, localization, or interaction with other molecules including proteins. However, given the therapeutic potential of *trans*-splicing ribozymes, further improvements of ribozymes through optimization and modification will make ribozymes applicable in a variety of clinical settings.

ACKNOWLEDGMENTS

This study was supported by the National Research Foundation of Korea (NRF) funded by the Ministry of Science and ICT (2017R1A2B2003368 and 2017M3A9C8031781) and by Business for Startup growth and technological development (TIPS Program) funded by Korea Small and Medium Business Administration in 2017 (Grant No. S2566697).

REFERENCES

1. Kaufmann KB, Buning H, Galy A, et al. Gene therapy on the move. *EMBO Mol Med.* 2013;5:1642–1661.
2. McCormick F. Cancer gene therapy: fringe or cutting edge? *Nat Rev Cancer.* 2001;1:130–141.
3. Collins M, Thrasher A. Gene therapy: progress and predictions. *Proc Biol Sci.* 2015;282: 20143003.

4. Naldini L. Gene therapy returns to centre stage. *Nature.* 2015;526:351–360.
5. Burnett JC, Rossi JJ. RNA-based therapeutics: current progress and future prospects. *Chem Biol.* 2012;19:60–71.
6. Scherer LJ, Rossi JJ. Approaches for the sequence-specific knockdown of mRNA. *Nat Biotechnol.* 2003;21:1457–1465.
7. Bennett CF, Swayze EE. RNA targeting therapeutics: molecular mechanisms of antisense oligonucleotides as a therapeutic platform. *Annu Rev Pharmacol Toxicol.* 2010;50:259–293.
8. Grimm D, Streetz KL, Jopling CL, et al. Fatality in mice due to oversaturation of cellular microRNA/short hairpin RNA pathways. *Nature.* 2006;441:537–541.
9. Gaj T, Gersbach CA, Barbas 3rd C. ZFN, TALEN, and CRISPR/Cas-based methods for genome engineering. *Trends Biotechnol.* 2013;31:397–405.
10. Barrangou R, Doudna JA. Applications of CRISPR technologies in research and beyond. *Nat Biotechnol.* 2016;34:933–941.
11. Hsu PD, Lander ES, Zhang F. Development and applications of CRISPR-Cas9 for genome engineering. *Cell.* 2014;157:1262–1278.
12. Pineda M, Moghadam F, Ebrahimkhani MR, et al. Engineered CRISPR systems for next generation gene therapies. *ACS Synth Biol.* 2017;6:1614–1626.
13. Cox DB, Platt RJ, Zhang F. Therapeutic genome editing: prospects and challenges. *Nat Med.* 2015;21:121–131.
14. Cornu TI, Mussolino C, Cathomen T. Refining strategies to translate genome editing to the clinic. *Nat Med.* 2017;23:415–423.
15. Yin H, Kauffman KJ, Anderson DG. Delivery technologies for genome editing. *Nat Rev Drug Discov.* 2017;16:387–399.
16. Doudna JA, Charpentier E. Genome editing. The new frontier of genome engineering with CRISPR-Cas9. *Science.* 2014;346:1258096.
17. Cong L, Ran FA, Cox D, et al. Multiplex genome engineering using CRISPR/Cas systems. *Science.* 2013;339:819–823.
18. Dahlman JE, Abudayyeh OO, Joung J, et al. Orthogonal gene knockout and activation with a catalytically active Cas9 nuclease. *Nat Biotechnol.* 2015;33:1159–1161.
19. Yang H, Wang H, Shivalila CS, et al. One-step generation of mice carrying reporter and conditional alleles by CRISPR/Cas-mediated genome engineering. *Cell.* 2013;154:1370–1379.
20. Kaczmarek JC, Kowalski PS, Anderson DG. Advances in the delivery of RNA therapeutics: from concept to clinical reality. *Genome Med.* 2017;9:60.
21. Moreno AM, Mali P. Therapeutic genome engineering via CRISPR-Cas systems. *Wiley Interdiscip Rev Syst Biol Med.* 2017;9:e1380.
22. Tycko J, Myer VE, Hsu PD. Methods for optimizing CRISPR-Cas9 genome editing specificity. *Mol Cell.* 2016;63:355–370.
23. Komor AC, Badran AH, Liu DR. CRISPR-based technologies for the manipulation of eukaryotic genomes. *Cell.* 2017;169:559.
24. Serganov A, Patel DJ. Ribozymes, riboswitches and beyond: regulation of gene expression without proteins. *Nat Rev Genet.* 2007;8:776–790.
25. Cech TR, Zaug AJ, Grabowski PJ. In vitro splicing of the ribosomal RNA precursor of Tetrahymena: involvement of a guanosine nucleotide in the excision of the intervening sequence. *Cell.* 1981;27:487–496.
26. Guerrier-Takada C, Gardiner K, Marsh T, et al. The RNA moiety of ribonuclease P is the catalytic subunit of the enzyme. *Cell.* 1983;35:849–857.
27. Kruger K, Grabowski PJ, Zaug AJ, et al. Self-splicing RNA: autoexcision and autocyclization of the ribosomal RNA intervening sequence of Tetrahymena. *Cell.* 1982;31:147–157.
28. Khan AU. Ribozyme: a clinical tool. *Clin Chim Acta.* 2006;367:20–27.

29. Lewin AS, Hauswirth WW. Ribozyme gene therapy: applications for molecular medicine. *Trends Mol Med*. 2001;7:221–228.
30. Puerta-Fernandez E, Romero-Lopez C, Barroso-delJesus A, et al. Ribozymes: recent advances in the development of RNA tools. *FEMS Microbiol Rev*. 2003;27:75–97.
31. Fiskaa T, Birgisdottir AB. RNA reprogramming and repair based on trans-splicing group I ribozymes. *N Biotechnol*. 2010;27:194–203.
32. Ban G, Jeong JS, Kim A, et al. Selective and efficient retardation of cancers expressing cytoskeleton-associated protein 2 by targeted RNA replacement. *Int J Cancer*. 2011;129:1018–1029.
33. Byun J, Lan N, Long M, et al. Efficient and specific repair of sickle beta-globin RNA by trans-splicing ribozymes. *RNA*. 2003;9:1254–1263.
34. Carter JR, Keith JH, Fraser TS, et al. Effective suppression of dengue virus using a novel group-I intron that induces apoptotic cell death upon infection through conditional expression of the Bax C-terminal domain. *Virol J*. 2014;11:111.
35. Jung HS, Kwon BS, Lee SW. Tumor-specific gene delivery using RNA-targeting Tetrahymena group I intron. *Biotechnol Lett*. 2005;27:567–574.
36. Kastanos E, Hjiantoniou E, Phylactou LA. Restoration of protein synthesis in pancreatic cancer cells by trans-splicing ribozymes. *Biochem Biophys Res Commun*. 2004;322: 930–934.
37. Kim SJ, Kim JH, Yang B, et al. Specific and efficient regression of cancers harboring KRAS mutation by targeted RNA replacement. *Mol Ther*. 2017;25:356–367.
38. Kwon BS, Jung HS, Song MS, et al. Specific regression of human cancer cells by ribozyme-mediated targeted replacement of tumor-specific transcript. *Mol Ther*. 2005;12:824–834.
39. Phylactou LA, Darrah C, Wood MJ. Ribozyme-mediated trans-splicing of a trinucleotide repeat. *Nat Genet*. 1998;18:378–381.
40. Rogers CS, Vanoye CG, Sullenger BA, et al. Functional repair of a mutant chloride channel using a trans-splicing ribozyme. *J Clin Invest*. 2002;110:1783–1789.
41. Ryu KJ, Kim JH, Lee SW. Ribozyme-mediated selective induction of new gene activity in hepatitis C virus internal ribosome entry site-expressing cells by targeted trans-splicing. *Mol Ther*. 2003;7:386–395.
42. Shin KS, Sullenger BA, Lee SW. Ribozyme-mediated induction of apoptosis in human cancer cells by targeted repair of mutant p53 RNA. *Mol Ther*. 2004;10:365–372.
43. Song MS, Lee SW. Cancer-selective induction of cytotoxicity by tissue-specific expression of targeted trans-splicing ribozyme. *FEBS Lett*. 2006;580:5033–5043.
44. Won YS, Lee SW. Selective regression of cancer cells expressing a splicing variant of AIMP2 through targeted RNA replacement by trans-splicing ribozyme. *J Biotechnol*. 2012;158:44–49.
45. Hong SH, Jeong JS, Lee YJ, et al. Molecular imaging of endogenous mRNA expression in a mouse tumor model by adenovirus harboring trans-splicing ribozyme. *FEBS Lett*. 2007;581:5396–5400.
46. Kim YH, Kim KT, Lee SJ, et al. Image-aided suicide gene therapy utilizing multifunctional hTERT-targeting adenovirus for clinical translation in hepatocellular carcinoma. *Theranostics*. 2016;6:357–368.
47. Kim YH, Moon JY, Kim EO, et al. Efficient targeting and tumor retardation effect of pancreatic adenocarcinoma up-regulated factor (PAUF)-specific RNA replacement in pancreatic cancer mouse model. *Cancer Lett*. 2014;344:223–231.
48. So MK, Gowrishankar G, Hasegawa S, et al. Imaging target mRNA and siRNA-mediated gene silencing in vivo with ribozyme-based reporters. *ChemBioChem*. 2008;9:2682–2691.
49. Michel F, Westhof E. Modelling of the three-dimensional architecture of group I catalytic introns based on comparative sequence analysis. *J Mol Biol*. 1990;216: 585–610.

50. Suh SO, Jones KG, Blackwell M. A group I intron in the nuclear small subunit rRNA gene of Cryptendoxyla hypophloia, an ascomycetous fungus: evidence for a new major class of group I introns. *J Mol Evol.* 1999;48:493–500.
51. Adams PL, Stahley MR, Gill ML, et al. Crystal structure of a group I intron splicing intermediate. *RNA.* 2004;10:1867–1887.
52. Adams PL, Stahley MR, Kosek AB, et al. Crystal structure of a self-splicing group I intron with both exons. *Nature.* 2004;430:45–50.
53. Golden BL, Kim H, Chase E. Crystal structure of a phage Twort group I ribozyme-product complex. *Nat Struct Mol Biol.* 2005;12:82–89.
54. Guo F, Gooding AR, Cech TR. Structure of the Tetrahymena ribozyme: base triple sandwich and metal ion at the active site. *Mol Cell.* 2004;16:351–362.
55. Stahley MR, Strobel SA. RNA splicing: group I intron crystal structures reveal the basis of splice site selection and metal ion catalysis. *Curr Opin Struct Biol.* 2006;16:319–326.
56. Vicens Q, Cech TR. Atomic level architecture of group I introns revealed. *Trends Biochem Sci.* 2006;31:41–51.
57. Been MD, Cech TR. One binding site determines sequence specificity of Tetrahymena pre-rRNA self-splicing, trans-splicing, and RNA enzyme activity. *Cell.* 1986;47:207–216.
58. Jackson S, Cannone J, Lee J, et al. Distribution of rRNA introns in the three-dimensional structure of the ribosome. *J Mol Biol.* 2002;323:35–52.
59. Burke JM, Esherick JS, Burfeind WR, et al. A 3' splice site-binding sequence in the catalytic core of a group I intron. *Nature.* 1990;344:80–82.
60. Waring RB, Davies RW. Assessment of a model for intron RNA secondary structure relevant to RNA self-splicing—a review. *Gene.* 1984;28:277–291.
61. Doherty EA, Doudna JA. The P4-P6 domain directs higher order folding of the Tetrahymena ribozyme core. *Biochemistry.* 1997;36:3159–3169.
62. Engelhardt MA, Doherty EA, Knitt DS, et al. The P5abc peripheral element facilitates preorganization of the tetrahymena group I ribozyme for catalysis. *Biochemistry.* 2000;39:2639–2651.
63. Joyce GF, van der Horst G, Inoue T. Catalytic activity is retained in the Tetrahymena group I intron despite removal of the large extension of element P5. *Nucleic Acids Res.* 1989;17:7879–7889.
64. Lehnert V, Jaeger L, Michel F, et al. New loop–loop tertiary interactions in self-splicing introns of subgroup IC and ID: a complete 3D model of the Tetrahymena thermophila ribozyme. *Chem Biol.* 1996;3:993–1009.
65. Cech TR. Self-splicing of group I introns. *Annu Rev Biochem.* 1990;59:543–568.
66. Cech TR. Catalytic RNA: structure and mechanism. *Biochem Soc Trans.* 1993;21: 229–234.
67. Cech TR, Bass BL. Biological catalysis by RNA. *Annu Rev Biochem.* 1986;55:599–629.
68. Golden BL, Gooding AR, Podell ER, et al. A preorganized active site in the crystal structure of the Tetrahymena ribozyme. *Science.* 1998;282:259–264.
69. Strobel SA, Ortoleva-Donnelly L, Ryder SP, et al. Complementary sets of non-canonical base pairs mediate RNA helix packing in the group I intron active site. *Nat Struct Biol.* 1998;5:60–66.
70. Kohler U, Ayre BG, Goodman HM, et al. Trans-splicing ribozymes for targeted gene delivery. *J Mol Biol.* 1999;285:1935–1950.
71. Sullenger BA, Cech TR. Ribozyme-mediated repair of defective mRNA by targeted, trans-splicing. *Nature.* 1994;371:619–622.
72. Johnson IM. RNA as a drug target: recent patents on the catalytic activity of trans-splicing ribozymes derived from group I intron RNA. *Recent Pat DNA Gene Seq.* 2010;4:17–28.
73. Lee SW, Jeong JS. Use of tumor-targeting trans-splicing ribozyme for cancer treatment. *Methods Mol Biol.* 2014;1103:83–95.

74. Long MB, Jones 3rd JP, Sullenger BA, et al. Ribozyme-mediated revision of RNA and DNA. *J Clin Invest*. 2003;112:312–318.
75. Jones JT, Lee SW, Sullenger BA. Tagging ribozyme reaction sites to follow trans-splicing in mammalian cells. *Nat Med*. 1996;2:643–648.
76. Campbell TB, Cech TR. Identification of ribozymes within a ribozyme library that efficiently cleave a long substrate RNA. *RNA*. 1995;1:598–609.
77. Ryu KJ, Lee SW. Identification of the most accessible sites to ribozymes on the hepatitis C virus internal ribosome entry site. *J Biochem Mol Biol*. 2003;36:538–544.
78. Einvik C, Fiskaa T, Lundblad EW, et al. Optimization and application of the group I ribozyme trans-splicing reaction. *Methods Mol Biol*. 2004;252:359–371.
79. Milligan JF, Groebe DR, Witherell GW, et al. Oligonucleotide synthesis using T7 RNA polymerase and synthetic DNA templates. *Nucleic Acids Res*. 1987;15:8783–8798.
80. Bustin SA. Absolute quantification of mRNA using real-time reverse transcription polymerase chain reaction assays. *J Mol Endocrinol*. 2000;25:169–193.
81. Meluzzi D, Olson KE, Dolan GF, et al. Computational prediction of efficient splice sites for trans-splicing ribozymes. *RNA*. 2012;18:590–602.
82. Cao Y, Woodson SA. Refolding of rRNA exons enhances dissociation of the Tetrahymena intron. *RNA*. 2000;6:1248–1256.
83. Hagen M, Cech TR. Self-splicing of the Tetrahymena intron from mRNA in mammalian cells. *EMBO J*. 1999;18:6491–6500.
84. Koduvayur SP, Woodson SA. Intracellular folding of the Tetrahymena group I intron depends on exon sequence and promoter choice. *RNA*. 2004;10:1526–1532.
85. Suh ER, Waring RB. Base pairing between the 3' exon and an internal guide sequence increases 3' splice site specificity in the Tetrahymena self-splicing rRNA intron. *Mol Cell Biol*. 1990;10:2960–2965.
86. Ayre BG, Kohler U, Turgeon R, et al. Optimization of trans-splicing ribozyme efficiency and specificity by in vivo genetic selection. *Nucleic Acids Res*. 2002;30:e141.
87. Jung HS, Lee SW. Ribozyme-mediated selective killing of cancer cells expressing carcinoembryonic antigen RNA by targeted trans-splicing. *Biochem Biophys Res Commun*. 2006;349:556–563.
88. Kim A, Ban G, Song MS, et al. Selective regression of cells expressing mouse cytoskeleton-associated protein 2 transcript by trans-splicing ribozyme. *Oligonucleotides*. 2007;17:95–103.
89. Guo F, Cech TR. In vivo selection of better self-splicing introns in Escherichia coli: the role of the P1 extension helix of the Tetrahymena intron. *RNA*. 2002;8:647–658.
90. Olson KE, Muller UF. An in vivo selection method to optimize trans-splicing ribozymes. *RNA*. 2012;18:581–589.
91. Porschke D. Elementary steps of base recognition and helix-coil transitions in nucleic acids. *Mol Biol Biochem Biophys*. 1977;24:191–218.
92. Beaudry AA, Joyce GF. Directed evolution of an RNA enzyme. *Science*. 1992;257: 635–641.
93. Hasegawa S, Choi JW, Rao J. Single-cell detection of trans-splicing ribozyme in vivo activity. *J Am Chem Soc*. 2004;126:7158–7159.
94. Hayden EJ, Ferrada E, Wagner A. Cryptic genetic variation promotes rapid evolutionary adaptation in an RNA enzyme. *Nature*. 2011;474:92–95.
95. Ohuchi SJ, Ikawa Y, Shiraishi H, et al. Artificial modules for enhancing rate constants of a group I intron ribozyme without a P4–P6 core element. *J Biol Chem*. 2004;79:540–546.
96. Treiber DK, Rook MS, Zarrinkar PP, et al. Kinetic intermediates trapped by native interactions in RNA folding. *Science*. 1998;279:1943–1946.
97. Amini ZN, Muller UF. Increased efficiency of evolved group I intron spliceozymes by decreased side product formation. *RNA*. 2015;21:1480–1489.

98. Olson KE, Dolan GF, Muller UF. In vivo evolution of a catalytic RNA couples trans-splicing to translation. *PLoS One*. 2014;9:e86473.
99. Zhao H, Giver L, Shao Z, et al. Molecular evolution by staggered extension process (StEP) in vitro recombination. *Nat Biotechnol*. 1998;16:258–261.
100. Muller UF. Design and experimental evolution of trans-splicing group I intron ribozymes. *Molecules*. 2017;22:75.
101. Dotson PP, Sinha J, Testa SM. A Pneumocystis carinii group I intron-derived ribozyme utilizes an endogenous guanosine as the first reaction step nucleophile in the trans excision-splicing reaction. *Biochemistry*. 2008;47:4780–4787.
102. Dotson PP, Johnson AK, Testa SM. Tetrahymena thermophila and Candida albicans group I intron-derived ribozymes can catalyze the trans-excision-splicing reaction. *Nucleic Acids Res*. 2008;36:5281–5289.
103. Fiskaa T, Lundblad EW, Henriksen JR, et al. RNA reprogramming of alpha-mannosidase mRNA sequences in vitro by myxomycete group IC1 and IE ribozymes. *FEBS J*. 2006;273:2789–2800.
104. Lan N, Howrey RP, Lee SW, et al. Ribozyme-mediated repair of sickle beta-globin mRNAs in erythrocyte precursors. *Science*. 1998;280:1593–1596.
105. Watanabe T, Sullenger BA. Induction of wild-type p53 activity in human cancer cells by ribozymes that repair mutant p53 transcripts. *Proc Natl Acad Sci U S A*. 2000;97: 8490–8494.
106. Powars DR, Weiss JN, Chan LS, et al. Is there a threshold level of fetal hemoglobin that ameliorates morbidity in sickle cell anemia? *Blood*. 1984;63:921–926.
107. Rucknagel DL, Sarniak SA, Whitten CF, et al. Fetal hemoglobin concentration predicts disease severity in children with sickle cell anemia. *Prog Clin Biol Res*. 1987;251:487–496.
108. Carter JR, Keith JH, Barde PV, et al. Targeting of highly conserved dengue virus sequences with anti-dengue virus trans-splicing group I introns. *BMC Mol Biol*. 2010;11:84.
109. Won YS, Lee SW. Targeted retardation of hepatocarcinoma cells by specific replacement of alpha-fetoprotein RNA. *J Biotechnol*. 2007;129:614–619.
110. Hong SH, Jeong JS, Lee YJ, et al. In vivo reprogramming of hTERT by trans-splicing ribozyme to target tumor cells. *Mol Ther*. 2008;16:74–80.
111. Jeong JS, Lee SW, Hong SH, et al. Antitumor effects of systemically delivered adenovirus harboring trans-splicing ribozyme in intrahepatic colon cancer mouse model. *Clin Cancer Res*. 2008;14:281–290.
112. Song MS, Jeong JS, Ban G, et al. Validation of tissue-specific promoter-driven tumor-targeting trans-splicing ribozyme system as a multifunctional cancer gene therapy device in vivo. *Cancer Gene Ther*. 2009;16:113–125.
113. Won YS, Lee SW. Cancer-specific induction of adenoviral E1A expression by group I intron-based trans-splicing ribozyme. *J Microbiol Biotechnol*. 2012;22:431–435.
114. Bukh J, Purcell RH, Miller RH. Sequence analysis of the 5' noncoding region of hepatitis C virus. *Proc Natl Acad Sci U S A*. 1992;89:4942–4946.
115. Lohmann V. Hepatitis C virus RNA replication. *Curr Top Microbiol Immunol*. 2013;369:167–198.
116. Usui K, Saijo Y, Narumi K, et al. N-terminal deletion augments the cell-death-inducing activity of BAX in adenoviral gene delivery to nonsmall cell lung cancers. *Oncogene*. 2003;22:2655–2663.
117. Song MS, Jeong JS, Cho KS, et al. Gene expression responses in vivo by human telomerase reverse transcriptase (hTERT)-targeting trans-splicing ribozyme. *Exp Mol Med*. 2007;39:722–732.
118. Ban G, Song MS, Lee SW. Cancer cell targeting with mouse TERT-specific group I intron of Tetrahymena thermophila. *J Microbiol Biotechnol*. 2009;19:1070–1076.

119. Kwon BS, Jeong JS, Won YS, et al. Intracellular efficacy of tumor-targeting group I intron-based trans-splicing ribozyme. *J Gene Med*. 2011;13:89–100.
120. Kim NW, Piatyszek MA, Prowse KR, et al. Specific association of human telomerase activity with immortal cells and cancer. *Science*. 1994;266:2011–2015.
121. Masutomi K, Hahn WC. Telomerase and tumorigenesis. *Cancer Lett*. 2003;194: 163–172.
122. Smith LL, Coller HA, Roberts JM. Telomerase modulates expression of growth-controlling genes and enhances cell proliferation. *Nat Cell Biol*. 2003;5:474–479.
123. Forsyth NR, Wright WE, Shay JW. Telomerase and differentiation in multicellular organisms: turn it off, turn it on, and turn it off again. *Differentiation*. 2002;69:188–197.
124. Shay JW, Wright WE. Telomerase: a target for cancer therapeutics. *Cancer Cell*. 2002;2:257–265.
125. Chen CZ, Li L, Lodish HF, et al. MicroRNAs modulate hematopoietic lineage differentiation. *Science*. 2004;303:83–86.
126. Ramkissoon SH, Mainwaring LA, Ogasawara Y, et al. Hematopoietic-specific microRNA expression in human cells. *Leuk Res*. 2006;30:643–647.
127. Georgantas 3rd RW, Hildreth R, Morisot S, et al. CD34+ hematopoietic stem-progenitor cell microRNA expression and function: a circuit diagram of differentiation control. *Proc Natl Acad Sci U S A*. 2007;104:2750–2755.
128. Won YS, Jeong JS, Kim SJ, et al. Targeted anticancer effect through microRNA-181a regulated tumor-specific hTERT replacement. *Cancer Lett*. 2015;356:918–928.
129. Kim J, Won R, Ban G, et al. Targeted regression of hepatocellular carcinoma by cancer-specific RNA replacement through microRNA regulation. *Sci Rep*. 2015;5: 12315.
130. Coulouarn C, Factor VM, Andersen JB, et al. Loss of miR-122 expression in liver cancer correlates with suppression of the hepatic phenotype and gain of metastatic properties. *Oncogene*. 2009;28:3526–3536.
131. Kim J, Jeong S, Kertsburg A, et al. Conditional and target-specific transgene induction through RNA replacement using an allosteric trans-splicing ribozyme. *ACS Chem Biol*. 2014;9:2491–2495.
132. Choi JW, Kim DG, Lee AE, et al. Cancer-associated splicing variant of tumor suppressor AIMP2/p38: pathological implication in tumorigenesis. *PLoS Genet*. 2011;7: e1001351.
133. Fernandez-Medarde A, Santos E. Ras in cancer and developmental diseases. *Genes Cancer*. 2011;2:344–358.
134. Okada T, Shah M, Higginbotham JN, et al. AV.TK-mediated killing of subcutaneous tumors in situ results in effective immunization against established secondary intracranial tumor deposits. *Gene Ther*. 2001;8:1315–1322.
135. Iida N, Nakamoto Y, Baba T, et al. Tumor cell apoptosis induces tumor-specific immunity in a CC chemokine receptor 1- and 5-dependent manner in mice. *J Leukoc Biol*. 2008;84:1001–1010.
136. Shin SP, Seo HH, Shin JH, et al. Adenovirus expressing both thymidine kinase and soluble PD1 enhances antitumor immunity by strengthening CD8 T-cell response. *Mol Ther*. 2013;21:688–695.

CHAPTER FIVE

Major 3′–5′ Exoribonucleases in the Metabolism of Coding and Non-coding RNA

Ricardo F. dos Santos, Ana P. Quendera, Sofia Boavida, André F. Seixas, Cecília M. Arraiano, José M. Andrade[1]
Instituto de Tecnologia Química e Biológica António Xavier, Universidade Nova de Lisboa, Oeiras, Portugal
[1]Corresponding author: e-mail address: andrade@itqb.unl.pt

Contents

1. Introduction 102
2. Polynucleotide Phosphorylase 107
   2.1 PNPase Function and Regulation 108
   2.2 PNPase Complexes 112
   2.3 PNPase Structure 113
   2.4 PNPase Substrates 115
3. RNase II 121
   3.1 RNase II Function and Regulation 122
   3.2 RNase II Structure 124
   3.3 RNase II Substrates 125
4. RNase R 126
   4.1 RNase R Function and Regulation 127
   4.2 RNase R Structure 129
   4.3 RNase R Substrates 132
   4.4 RNase II/RNase R in Eukaryotes 135
   4.5 RNase II/RNase R Family Members in Pathogenesis and Disease 136
5. Concluding Remarks 138
Acknowledgments 140
References 140

Abstract

3′–5′ exoribonucleases are key enzymes in the degradation of superfluous or aberrant RNAs and in the maturation of precursor RNAs into their functional forms. The major bacterial 3′–5′ exoribonucleases responsible for both these activities are PNPase, RNase II and RNase R. These enzymes are of ancient nature with widespread distribution. In eukaryotes, PNPase and RNase II/RNase R enzymes can be found in the cytosol and in mitochondria and chloroplasts; RNase II/RNase R-like enzymes are also found in the nucleus. Humans express one PNPase (PNPT1) and three RNase II/RNase R family

Progress in Molecular Biology and Translational Science, Volume 159
ISSN 1877-1173
https://doi.org/10.1016/bs.pmbts.2018.07.005

members (Dis3, Dis3L and Dis3L2). These enzymes take part in a multitude of RNA surveillance mechanisms that are critical for translation accuracy. Although active against a wide range of both coding and non-coding RNAs, the different 3′–5′ exoribonucleases exhibit distinct substrate affinities. The latest studies on these RNA degradative enzymes have contributed to the identification of additional homologue proteins, the uncovering of novel RNA degradation pathways, and to a better comprehension of several disease-related processes and response to stress, amongst many other exciting findings. Here, we provide a comprehensive and up-to-date overview on the function, structure, regulation and substrate preference of the key 3′–5′ exoribonucleases involved in RNA metabolism.

1. INTRODUCTION

The level at which any RNA is expressed is determined by its rate of transcription and its rate of degradation. Cells are constantly producing new RNAs while discarding old or defective transcripts in order to regulate gene expression. Coding RNA/messenger RNA (mRNA) decay is essential because of its role in destruction of defective mRNAs producing aberrant proteins, recycling of ribonucleotides, and adaptation to changes in the environment. Moreover, RNA transcripts usually have to be processed and/or modified during their maturation. The structure and chemically reactive nature of RNA make this molecule more unstable than DNA. However, RNAs are not all degraded at the same rate or in the same manner. Consequently, cells must control precisely the various enzymatic activities involved in RNA degradation and processing to avoid elimination of important RNA molecules. The enzymes that cleave and trim newly made transcripts to mature them and to rapidly destroy unnecessary cellular RNAs are called ribonucleases. These enzymes are described as endoribonucleases, if they cleave RNA molecules internally, or exoribonucleases if they attack the ends of the RNAs.

In prokaryotes, RNA decay is primarily initiated by endonucleolytic cleavages, followed by the action of exoribonucleases that rapidly eliminate the intermediary products.[1,2] However, exoribonucleases can also act on primary transcripts.[3] Gram-negative bacteria do not encode 5′–3′ exoribonucleases, which seem to be less abundant than 3′–5′ exoribonucleases.[4] Bacterial RNases have been most extensively studied in *Escherichia coli*, which encodes about 20 RNases, of which 8 are 3′–5′ exoribonucleases.[5] In this review we will focus on the major 3′–5′ exoribonucleases, PNPase, RNase II and RNase R, which exhibit a broader role in RNA metabolism. RNases BN, D, PH and T are mostly associated with RNA processing and oligoribonuclease acts as the "finishing" enzyme degrading the short oligoribonucleotides (2–4 nucleotides in length) released by the action of other ribonucleases. These more specialized 3′–5′ exoribonucleases are not addressed here but are reviewed in Refs. 3,5.

PNPase (encoded by *pnp*) is a 3′–5′ phosphorolytic enzyme member of PDX family of enzymes. RNase II (*rnb*) and RNase R (*rnr*) are paralogue proteins that belong to the RNase II family of enzymes that hydrolytically degrade RNA in the 3′ to 5′ direction in a processive manner.[6] These enzymes have homologues in all domains of life (Figs. 1 and 2). PNPase differs from RNase II and RNase R not only in the mode of action but also in its structure. The homotrimeric PNPase is assembled into a ring-like structure with a central channel comprising six RNase PH-like domains.[11] PNPase homologues are found in eukaryotic organelles, with a similar domain organization and structure. Moreover, this overall organization is

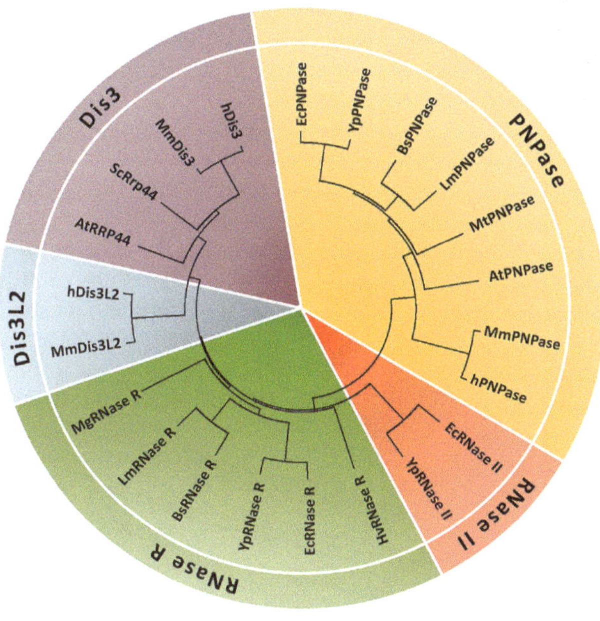

Fig. 1 Phylogenetic tree of the major 3′–5′ exoribonucleases showing the relationship between the different proteins across representative organisms. The evolutionary analysis was conducted in MEGA X software using the neighbor-joining method and sequence alignment was performed with MUSCLE algorithm.[7] Organisms are as follows: Ec, *Escherichia coli*; Yp, *Yersinia pestis*; Bs, *Bacillus subtilis*; Lm, *Listeria monocytogenes*; Mt, *Mycobacterium tuberculosis*; Mg, *Mycoplasma genitalium*; Hv, *Haloferax volcanii*; At, *Arabidopsis thaliana*; Sc, *Saccharomyces cerevisiae*; Mm, *Mus musculus*; h, *Homo sapiens*.

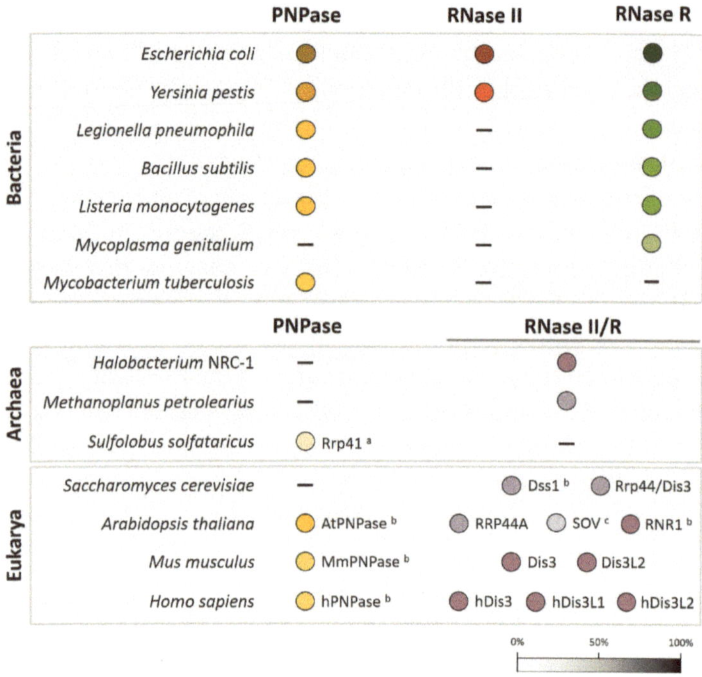

Fig. 2 Comparative analysis of 3′–5′ exoribonuclease homologues among representative organisms. The amino acidic sequence conservation is represented, when applicable, by filled circles and the brightness/darkness of each circle depicts the degree of conservation (scale at the bottom). The *Escherichia coli* exoribonucleases were used as reference for comparison with other species using the co-occurrence analysis of the STRING database.[8] Protein names are shown at the right of the corresponding circle in cases where the name of the homologue is different from the reference enzymes in *E. coli*. [a]The homology analysis between the PNPase from *E. coli* and the Rrp41 from *S. solfataricus* is represented. However, the latter is not an actual PNPase homologue (as in eukaryotic PNPase) but rather a phosphorolytic exoribonuclease that belongs to the same PDX protein family.[9] [b]These homologues, although nuclear encoded, localize to the mitochondria of eukaryotic cells and to both the mitochondria and chloroplasts of plant cells (*A. thaliana*). [c]SOV is a Dis3L2-like cytoplasmatic enzyme that affects chloroplasts gene expression.[10]

similar to those of archaeal and eukaryotic multiprotein exosome core complexes, indicating evolutionary links in structure and function between PNPase and exosomes. The RNase II/RNase R enzymes are characterized by the presence of a well-conserved RNB domain that has a unique αβ-fold.[12] The RNB catalytic domain was named after the *E. coli rnb* gene that encodes ribonuclease B, widely known as ribonuclease II (RNase II)

(Fig. 3). RNase II from *E. coli* is the prototype of this family that also includes bacterial RNase R and eukaryotic Rrp44/Dis3 proteins. Some organisms like *Saccharomyces cerevisiae* may express only one member, while others may express up to three paralogues. In Bacteria, the γ division *Proteobacteria* often contain both RNase II and RNase R homologues. However, outside of this lineage, the single family member present more closely resembles RNase R than RNase II.[6] In Archaea and Eukarya, classification of the homologues members follows the generic designation of RNase II-like or RNase II/RNase R enzymes (Figs. 1 and 2). These enzymes are also important in several stress responses and have been associated to pathogenesis.[3]

PNPase, RNase II and RNase R may display some functional redundancy, as they share common substrates. However, they do not show the same degree of ability to compensate each other's activity. Cells must express PNPase together with at least one of the hydrolytic enzymes RNase II or RNase R, as double mutants *pnp rnb* or *pnp rnr* are lethal.[13,14] However, the double *rnb rnr* strain is viable. This highlights the importance of PNPase in cell biology. These 3′–5′ enzymes are active against a wide range of substrates, including coding (mRNA) and non-coding RNAs (ribosomal RNA/rRNA, transfer RNA/tRNA, small RNA/sRNA). PNPase and RNase II are mainly involved in mRNA decay but also participate in tRNA and rRNA maturation and degradation.[13,15,16] PNPase is the main enzyme affecting small non-coding RNAs stability.[17–19] RNase R is more important in quality control mechanisms that assure the degradation of aberrant rRNA and tRNA molecules, though it has also been related to mRNA decay.[3,16,20,21] A common substrate requirement for all the major 3′–5′ exoribonucleases is the need for a linear overhang in the 3′ end of the RNA molecule to promote binding of the enzyme. Hence, the degradation of RNA molecules by PNPase, RNase II and RNase R is promoted by polyadenylation, the addition of poly(A) stretches at the 3′ end of the RNA.[22]

Here we present a comprehensive overview of the major bacterial 3′–5′ exoribonucleases that are involved in the degradation and processing of RNAs. We have focused in the prokaryotic enzymes, but we also make evolutionary comparisons with their eukaryotic counterparts. These enzymes are conserved and widely distributed, which argues in favor of their critical functions in cell biology. In fact, they are important players in the metabolism of coding and non-coding RNA.

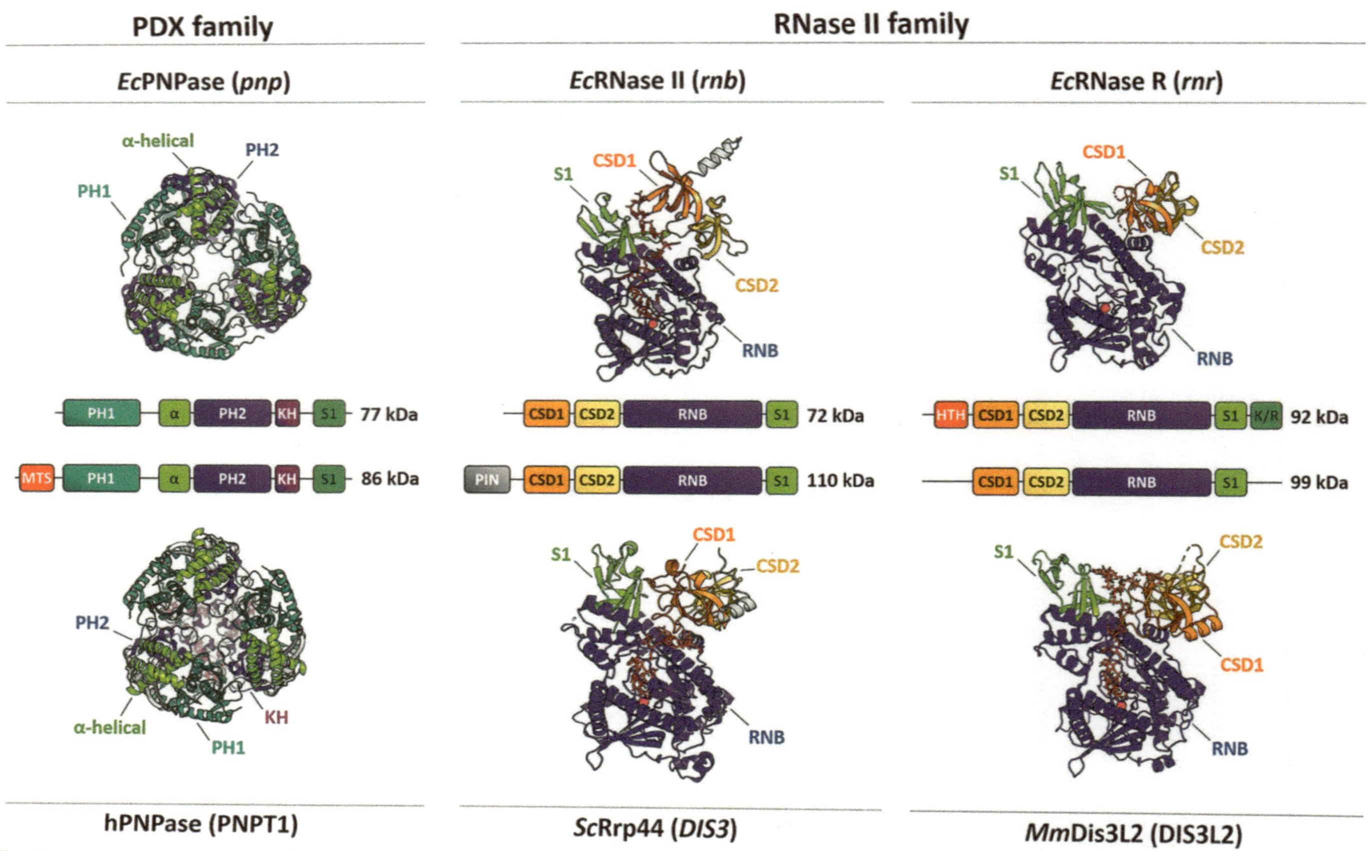

Fig. 3 See legend on opposite page.

2. POLYNUCLEOTIDE PHOSPHORYLASE

Polynucleotide phosphorylase (PNPase) is one of the most remarkable enzymes involved in RNA metabolism. This enzyme is unique amongst exoribonucleases as it presents dual activity: it works either as a 3′–5′ exoribonuclease in RNA degradation or as a polymerase in RNA synthesis, depending on the inorganic phosphate (Pi) concentration available.[23] The seminal work by Marianne Grunberg-Manago led to the discovery of PNPase as the first enzyme able to synthesize RNA *in vitro*.[24] This activity was essential for elucidation of the genetic code, leading to Severo Ochoa's Nobel Prize in Physiology or Medicine in 1959. Nonetheless, today it is widely recognized that its main activity in cells is RNA degradation. In recent years, PNPase has become a key enzyme in the regulation of non-coding RNAs across species. PNPase is widely conserved from bacteria

Fig. 3 Schematic representation of the domain organization and crystal structures of representative members of the PDX and RNase II protein families. The name of each protein along with the corresponding gene between brackets is shown next to each crystal structure. The domain organization followed by the approximate molecular weight of each prokaryotic and eukaryotic member (*top* and *bottom panels*, respectively) is aligned and displayed on top of each other for easy comparison. The specific domains present in each 3D crystal structure are also depicted. The Mg^{2+} cofactor is represented as a sphere within the catalytic centers, when applicable. The PDX family members *Ec*PNPase and hPNPase primary structure contains two RNase PH folds (PH1 and PH2) linked by an α-helical region (*α*), followed by two RNA binding domains, KH and S1. The hPNPase exhibits an additional N-terminal MTS domain containing the signal peptide for the mitochondrial localization. Both PNPases are trimers that assemble in a doughnut-like shape and are displayed here along the threefold rotation axis. All representative members of the RNase II family exhibit a similar domain arrangement: a central catalytic RNB domain flanked at the N-terminus by two cold shock RNA binding domains (CSD1 and CSD2) and at the C-terminus by an RNA binding S1 domain. Additionally, *Sc*Rrp44 exhibits an N-terminal PIN domain, whereas the *Ec*RNase R comprises two extra domains: a helix-turn-helix domain (HTH) at the N-terminus and a K/R rich patch at the C-terminus. The RNase II family members are displayed here in a front view of the RNB domain capped by the flanking RNA binding domains. Structures containing bound RNA molecules are displayed to allow visualization catalytic feeding channels and substrate-interacting domains. *Ec*RNase R structure is the sole exception since it has not yet been cocrystallized with a bound substrate. Organisms and PDB identification codes are the following: *Ec*PNPase, *Escherichia coli* PNPase (PDBID: 3GCM); hPNPase, *Homo sapiens* PNPase (PDBID: 3U1K); *Ec*RNase II, *Escherichia coli* RNase II (PDBID: 2IX1); *Sc*Rrp44, *Saccharomyces cerevisiae* Rrp44 (PDBID: 2VNU); *Ec*RNase R, *Escherichia coli* RNase R (PDBID: 5XGU); *Mm*Dis3L2, *Mus musculus* Dis3L2 (PDBID: 4PMW).

to plants and metazoans.[6,25] Studies on this enzyme have contributed enormously, and still do, to a better understanding of RNA genetics and its regulation.

2.1 PNPase Function and Regulation

PNPase is a member of the PDX family, along with RNase PH found in bacteria, and the core of the exosome found in archaea and eukaryotes.[6] The main activity of PNPase in cells is the degradation of RNA. However, depending on inorganic phosphate concentration, this enzyme can act as a synthetic enzyme leading to polymerization of single-stranded RNA (ssRNA) from nucleoside diphosphates (dNDPs). When acting on RNA, Mg^{2+} is an essential cofactor.[3] Another reaction less explored is its ability to exchange the β-phosphate group of dNDPs and free orthophosphate.[26] Nevertheless, PNPase is also involved in single-stranded DNA (ssDNA) degradation and DNA repair pathways, reactions that are mostly dependent on the presence of Mn^{2+} ions.[27] Therefore, PNPase is a multifunctional protein active against a wide range of substrates.

2.1.1 PNPase Activity on RNA

PNPase catalyzes the processive 3′–5′ phosphorolytic degradation of RNA, releasing nucleoside diphosphates in the presence of high inorganic phosphate concentrations. To be able to bind to an RNA substrate, the RNA molecule must have a minimal 3′ overhang of 7–10 unpaired nucleotides (nts).[28,29] A wide-genome analysis showed that PNPase affects the stability of more transcripts than any other 3′–5′ exoribonuclease.[30] Deletion of PNPase resulted in the upregulation of 59% of transcripts (compared to only 29% and 41% in deletion strains of RNase II and RNase R, respectively). Moreover, many of these transcripts were stable RNAs (tRNAs, rRNAs and sRNAs). In fact, PNPase seems to be particularly relevant for stability control of non-coding RNAs.[17,19,31] PNPase is involved not only in the degradation but also in the processing and maturation of a variety of substrates (see Section 2.4). PNPase acts in a processive way and is sequence independent, though its action on folded RNAs can be stimulated by 3′ polyadenylation.[28,32,33] However, the processive degradation of RNA by PNPase is blocked by double-stranded RNA (dsRNA) structures.[33] To overcome such physical blocks on RNA, PNPase can associate with other proteins in a variety of multiprotein complexes (see Section 2.2).

At low inorganic phosphate concentrations, however, PNPase catalyzes the polymerization of ssRNA from dNDPs in a template-independent manner.[34,35] Firstly identified *in vitro*, this reaction was further observed to occur *in vivo* as PNPase was shown to add polyribonucleotide tails in *E. coli* strains lacking PAP I, the major polyadenylating enzyme.[23] Contrarily to the homopolymeric adenosine tails added by PAP I, the PNPase-biosynthesized tails are heteropolymeric containing all four nucleotides.[36] Rho-dependent transcription terminators were suggested to be modified by the polymerase activity of PNPase, while Rho-independent transcription terminators act as polyadenylation signals for PAP I.[37] Although PAP I is the major polyadenylating enzyme in *E. coli*, responsible for modification of more than 90% of the transcripts in exponentially growing bacteria, PNPase seems to be the main polyadenylating enzyme in *Streptomyces coelicolor*, spinach chloroplasts and cyanobacteria.[38–41] The archaeal exosome, structurally very similar to PNPase, has also been demonstrated to be responsible for the addition of heteropolymeric tails in the archaeon *Sulfolobus solfataricus*.[9] Hence, rounds of addition of heteropolymeric tails and exoribonucleolytic digestion by PNPase can promote the degradation of RNA in different organisms.

2.1.2 PNPase Activity on DNA

Additionally, PNPase has phosphorolytic activity on DNA and catalyzes template-independent synthesis of ssDNA from dNDPs in the presence of Mn^{2+} or Fe^{3+} ions.[42,43] Its ability to synthesize oligodeoxyribonucleotides was discovered in the 1950s, but there was no evidence that it had a role *in vivo*.[24] Several pieces of evidence linking PNPase and DNA metabolism have now emerged. Firstly, it was shown that *Bacillus subtilis* PNPase can play an important role in DNA repair pathways by acting in the degradation of ssDNA, which occurs in the presence of Mn^{2+} and at low levels of Pi. In contrast, in the presence of Mg^{2+} and at high levels of Pi, this enzyme degrades RNA but not DNA.[27] Secondly, PNPase was found to co-purify with RecN, a key protein involved in DNA damage recognition and joining DNA ends after double-strand breaks (DSBs). RecN stimulates Mn^{2+}-dependent PNPase polymerization on 3′-OH ends.[44] The polymeric and ssDNA degrading activities were both observed in *B. subtilis* and *E. coli*, which suggests conserved functions in two evolutionarily distant bacteria. PNPase is probably involved in homologous recombination (HR) and non-homologous end joining (NHEJ) pathways of DSB repair. It was suggested that PNPase either converts non-ligating breaks to clean ends sealable by DNA ligase or adds single-stranded 3′-tails without template that can then influence the repair pathway choice.[27,44] Thirdly, deletion of the

E. coli PNPase-coding gene partially suppressed a couple of mutator phenotypes, indicating a possible implication in spontaneous mutagenesis during replication.[45] It was suggested that the concentration of available nucleotides and the rate of base misincorporation during replication is affected by PNPase-dependent phosphorolysis since there are more ribonucleotide diphosphates available for building dNTPs.[46]

2.1.3 Regulation of PNPase Activity

The RNA chaperone Hfq is a mediator of sRNA-mRNA interactions that has been shown to display other relevant functions in the cell.[47] Hfq was also found to regulate PNPase activity. The biosynthetic activity of PNPase in the *hfq* single mutant is enhanced and it becomes the primary polynucleotide polymerase, adding heteropolymeric tails almost exclusively to 3′ truncated mRNAs.[48] Previous observations have hinted at a relationship between PNPase activity and cellular metabolism. For instance, PNPase can be regulated by nucleotides; the binding of ATP to PNPase results in inhibition of both the phosphorolytic and the polymeric activities of this exoribonuclease.[49] On the other hand, c-di-GMP activates PNPase activity in a reaction that is dependent on oxygen and controlled by its association with DosC and DosP direct oxygen sensor proteins.[50] Earlier crystallographic studies of *E. coli* PNPase revealed the presence of citrate from the crystallization buffer at the active site.[51] When the putative role of citrate in PNPase activity was further explored, it was observed that its enzymatic activity was inhibited in the presence of magnesium-chelated citrate.[52] Interestingly, human PNPase (hPNPase) regulates RNA import to the mitochondria where the Krebs cycle takes place and citrate is present at low concentrations.[53] Overall, the data show that PNPase activity is dependent on the metabolic state of the cell. Curiously, hPNPase and the archaeal exosome complex from *S. solfataricus* activities are also inhibited by the presence of citrate, suggesting a conserved link between RNA-degrading activity and cellular metabolism throughout evolution.[54]

2.1.4 Regulation of PNPase Expression

PNPase is encoded by the *pnp* gene, which is part of a two-gene operon with the upstream *rpsO* gene that encodes the ribosomal protein S15. The transcription of *pnp* is mainly driven from two promoters, one upstream of *rpsO* and one in the intergenic region.[55] PNPase expression is regulated at the post-transcriptional level by many factors. Two of the most important and the first to be identified were PNPase itself and RNase III.[56] A stem-loop structure found upstream of the ribosome binding site (RBS) sequence in the 5′UTR makes the *pnp* mRNA very stable. In normal conditions, RNase III cleaves at this double-stranded region, originating an RNA helix with a protruding 3′ end that is then preferentially removed by PNPase. The processed transcript is then rapidly degraded.[56,57] PNPase is thus autogenously regulated and the stability of the *pnp* mRNA is inversely correlated with the amount of active PNPase in the cell.[58] Later, it was shown that PNPase can act as a translational repressor of its own expression independently of RNase III cleavage.[59] In the absence of RNase III, PNPase can inhibit its own translation possibly by competing for *pnp* primary transcript with the ribosomal protein S1. Polyadenylation can indirectly affect PNPase autoregulation and thus PNPase levels. The higher the levels of polyadenylated transcripts are, the more likely PNPase binds to them instead of the 5′ end of its own transcripts.[60] Interestingly, RNase II also regulates PNPase expression because in the absence of RNase II, there is an overexpression of PNPase.[61]

The RNA-binding protein CsrA was also found to repress PNPase translation.[62] CsrA protein is a component of the Csr system that is widely involved in regulation of virulence, motility, biofilm formation, quorum sensing, carbon metabolism and stringent response.[63] After the *pnp* 5′UTR stem-loop is processed by RNase III and PNPase, CsrA binds to two sites of the transcript, where one CsrA site overlaps the *pnp* Shine-Dalgarno sequence, preventing ribosome binding and leading to the repression of *pnp* translation.[62] CsrA-mediated repression of PNPase was observed throughout growth, but it was greater during stationary phase. Being part of the Csr system, CsrA is negatively regulated by sRNAs CsrB and CsrC, which suggests that alterations in CsrB/C levels would translate into alterations in PNPase levels.

PNPase levels are induced in cells exposed to a temperature downshift. At optimal temperature, PNPase is not essential to *E. coli*, unless either RNase II or RNase R are missing.[13,14] Interestingly, a strain lacking both RNase II and RNase R is viable, meaning that PNPase can replace their activity in those conditions.[21] Nonetheless, at low temperatures, PNPase is indispensable for *E. coli* growth,[64,65] and has been identified as a cold shock protein.[66] During the cold acclimation phase there is a 10-fold increase in *pnp* transcripts[65] and a 2-fold increase in PNPase protein level.[67] This increase in transcripts appears to be due mainly to their stabilization.[65,67] However, RNase III processing of *pnp* transcripts is not affected by the temperature downshift. Since the *pnp* mRNA is not efficiently translated during cold acclimation, the stabilization of the transcripts appears to compensate for this poor translation to maintain or increase the protein levels.[65] After adaptation to cold, *pnp* mRNA levels decrease most probably due to RNase R activity.[68]

Proteins are not the only regulators of PNPase expression. *In vitro* assays have showed that the small non-coding SraG RNA (located upstream of the *pnp* gene) binds with *pnp* mRNA and induces new sites that can be recognized and cleaved by RNase III within the *pnp* mRNA region base-paired with SraG.[69] Furthermore, the SraG-*pnp* mRNA basepairing prevents the *pnp* translational initiation *in vitro* because it blocks the 30S subunit docking onto the RBS sequence of the transcript. Moreover, SraG-dependent regulation of PNPase expression was found to occur either in the presence or absence of RNase III.[69] It is proposed that such regulation can be relevant after induction of the stringent response, which results in the 2-fold increase in the levels of the regulatory non-coding RNA SraG.

2.2 PNPase Complexes

PNPase is sensitive to RNA secondary structures and the processive degradation of RNA by PNPase is blocked by double-stranded RNA regions.[33] PNPase can associate with different proteins, which form multiprotein complexes helpful to overcome this physical challenge. The most well-known PNPase-based complex is the degradosome, a multiprotein complex that is mainly composed of the endoribonuclease RNase E, which provides the scaffold for the complex, the DEAD-box RhlB helicase and the glycolytic enzyme enolase.[28,70,71] The degradosome composition is dynamic and varies in response to environmental changes. For instance, at cold temperatures the helicase RhlB is replaced by the helicase CsdA.[72]

The degradosome complex (or variations of it) is found in many different organisms. In the Gram-negative *Caulobacter crescentus*, an RNA degradosome complex was also identified.[73] It is composed by PNPase, RNase E as the scaffold, a DEAD-box helicase, and, surprisingly, the Krebs cycle enzyme aconitase. This is another proof of the connection between RNA degradation and cellular metabolism, as mentioned above. Based on bacterial two-hybrid system and co-purification assays, an RNA degradosome has also been proposed to exist in the Gram-positive *B. subtilis*. In the absence of RNase E, it is RNase Y that provides the scaffold for the complex, which includes PNPase, RNase J1, the DEAD-box helicase CshA and two glycolytic enzymes, enolase and phosphofructokinase.[74–76] Nevertheless, *B. subtilis* degradosome partners have not yet been purified in complex. Furthermore, few components colocalized: RNase Y is located at the membrane, RNase J1/J2 and CshA are close to the RNA bulk in the cytoplasm, while PNPase and the glycolytic enzymes are uniformly dispersed in the cytoplasm.[77] These results do not exclude completely the interactions between the components but suggest that these interactions could be only transitory.

Like in *B. subtilis*, the components of the RNA degradosome are the same in *Staphylococcus aureus*. RNase Y interacts with the RNA helicase CshA and enolase, although there is no confirmed interaction between PNPase and RNase Y.[78] Recently, it has been suggested that RnpA, a subunit of RNase P, could also have a role in this degradosome.[79] There is competition between PNPase and enolase for RnpA, though PNPase catalytic activity was unaltered when bound to RnpA and enolase activity was slightly upregulated.[79] On the other hand, in *Staphylococcus epidermis*, it was observed that PNPase forms a complex with RNase J1 and RNase J2.[80] These interactions have not changed either catalytic activities of PNPase, which could mean that they could be more important to increase degradation of structured RNA in certain environmental or cellular contexts.

PNPase is able to associate with RhlB independently of the degradosome, and this complex was found to be functional for the *in vitro* degradation of dsRNAs.[81,82] This PNPase–RhlB interaction was later observed to be important for cysteine regulation *in vivo*, since its disruption led to the increase of intracellular content of cysteine and enhanced anti-oxidative resistance.[83] Additionally, PNPase can associate with Hfq and/or PAP I. Such complexes are important for the poly(A)-dependent decay of RNAs. Moreover, as previously stated, Hfq stimulates PNPase polymerase activity.[48]

2.3 PNPase Structure

Pure PNPase was isolated from *E. coli* as a homotrimer of 78 kDa subunits.[84] The X-ray crystal structures of *E. coli*, *Streptomyces antibioticus* and *C. crescentus* PNPase reveal a homotrimeric subunit organization with a ring-like architecture (Fig. 3).[11,51,85,86] Each monomer exhibits a five-domain arrangement: at the N-terminus, two RNase PH domains (first and second core domains) are linked by an α-helical domain[87]; at the C-terminal end, two RNA-binding domains named KH and S1[88] (Fig. 3). In the quaternary structure, the KH and S1 domains are found together in one face of the trimer, while the active site (RNase PH domains) is in the opposite side. The three subunits associate via trimerization interfaces of the core domains, forming a central channel where catalysis occurs. The existence of conserved basic residues in the neck region, along with a proper constriction in the channel, are pivotal in capturing RNA for processive degradation.[85] Two constriction points have been identified in the channel, and the structure of PNPase in complex with RNA clearly indicates that the pathway followed by the RNA molecule is along the central pore in the direction of the active site.[11,51,85] The dynamic translocation of RNA by the enzyme depends on the conformational changes that occur in the opening at the central channel and its neighboring regions.[51] Kinetic studies have established that the presence of distinct and separated RNA-binding sites causes, in part, the processivity of PNPase. The trimer channel probably contributes both to processivity and to the regulation of PNPase activity by RNA structural elements.[11]

The first core RNase PH domain has a conserved FFRR loop interacting with RNA, approximately 20 Å away from the putative catalytic site in the second core.[89] Two *E. coli* PNPase mutants in the FFRR loop (Arg79Ala and Arg80Ala) showed a great increase in K_m for ADP/Pi binding. Moreover, the first core domain was suggested to be involved in the guanosine pentaphosphate (pppGpp) synthetase activity of *S. antibioticus* PNPase, absent in the majority of PNPases.[11] The second core RNase PH domain binds tungstate, a phosphate analogue, which suggests that the catalytic center is located here. This region may also have been separately adapted as a second active site in PNPases in general.[11] Most of the conserved residues are clustered in the second core domain (which contains the putative catalytic center) and in a small part of the first core domain. Only a few are in the KH or S1 RNA-binding domains. Mutations in amino acids located around the tungstate binding site abolish or severely decrease all catalytic activities of the enzyme, suggesting that these mutations affect the catalytic site directly.[90] *E. coli* PNPase crystals obtained in the presence of Mn^{2+} showed that the co-factor is coordinated by the conserved residues Asp486, Asp492 and Lys494.[51] Moreover, mutation of Asp492 eliminates both phosphorolytic and polymeric activities.[90]

The α-helical domains, which connect the first and second core domains, have been implicated in the catalytic activity of *E. coli* PNPase,[91] though they are the least conserved domains.[25,90] In *E. coli* PNPase, mutating adjacent residues the α-helical domain seems to affect the catalytic activity.[91] Analysis of the *S. antibioticus* structure demonstrated that helix α1 faces the putative phosphate binding site and could thus be a component of the catalytic center.[11,91] These results along with studies in spinach chloroplast and human PNPases[92,93] suggest that the catalytic site of *E. coli* PNPase is probably composed of structural elements in both the first and second core domains (Fig. 3).

On the other hand, the RNA-binding domains have limited effects on phosphorolysis, as PNPase mutants lacking either domains are catalytically active. RNA electrophoretic mobility shift assays showed that both KH and S1 domains are required for a proper binding[94] and for PNPase autoregulation.[95] Interestingly, PNPase truncated in these two domains can still bind RNA, though with weaker affinities, which confirms that the catalytic core has intrinsic RNA-binding activity. Additionally, the S1 and KH domains are required for substrate release from the catalytic center.[96] A two-step model was proposed to explain this, based on the indirect facilitation of PNPase activity by both KH and S1 domains, allowing substrate binding and product release. The model proposes that there are two RNA-binding surfaces on each PNPase monomer: one in the core catalytic domain and another formed by KH and S1 domains. First, it is suggested that ssRNA molecules interact weakly with one of the tandem KH–S1 domains. Then this bound ssRNA interacts strongly with the catalytic site, where it is subjected to phosphorolysis, until a stem-loop appears and the enzyme stalls because the substrate remains bound to the core RNA-binding surface. Finally, another molecule could bind to other unoccupied tandem KH–S1 domain, migrate to the core RNA-binding site and allow the displacement of the stalled molecule. Hence, truncated ΔKH–S1 proteins are much less efficient in product release after catalysis.[96]

2.4 PNPase Substrates

2.4.1 PNPase in the Metabolism of Coding RNA

Whole-genome transcriptomic studies identified that PNPase is the exoribonuclease that most significantly affects *E. coli* transcripts.[30,97,98] In a recent RNA-seq study, a total of 226 differentially expressed transcripts were reported in the deletion *pnp* mutant strain.[30] Not surprisingly, PNPase was found to affect many different pathways in the cell. However, this study highlighted the importance of PNPase in contributing to cell motility in *E. coli*. The *pnp* mutant was significantly less motile and formed less biofilms than the parental strain, although the genes under PNPase regulation responsible for these phenotypes were not totally clear.[30] Nonetheless, these findings in *E. coli* are in agreement with reports in *B. subtilis*. Inactivation of PNPase in *B. subtilis* also triggered a deficient swarming motility phenotype.[99,100] This was shown to be dependent on the PNPase-mediated degradation of the *slrA* mRNA. SlrA protein controls the levels of *fla/che* operon and, indirectly, *sigD* regulon, controlling the major determinants of cell motility.[100] Hence, regulation of RNA turnover is a major determinant of motility gene expression and help explain PNPase impact on the virulence of different pathogenic bacteria.

PNPase is key in a surveillance mechanism that removes oxidized RNA. The exposure of cells to oxidative stress results in the formation of an oxidized form of guanine, the 8-oxoguanine, which is particularly harmful as it can be used as substrate by RNA polymerase. PNPase can discriminate 8-oxoguanine-containing RNA from normal RNA. It binds specifically to 8-oxoguanine-containing RNAs and removes these faulty molecules from the translational machinery.[101] As consequence, PNPase was found to protect both *E. coli* and HeLa cells from oxidative stress.[83,102,103] This constitutes another example that PNPase contributes to a high fidelity of translation.

2.4.2 PNPase in the Metabolism of Non-coding RNA

In addition to its role in mRNA turnover, PNPase is involved not only in the degradation but also in the processing of a variety of stable RNAs (tRNAs, rRNAs and sRNAs) in different organisms. In *E. coli*, all tRNAs are encoded with extra nucleotides at both 5′ and 3′ ends. Many of these transcripts terminate in a Rho-independent way, generating a stem-loop at the 3′ end. PNPase is involved in the maturation and repair of the tRNAs 3′-terminus.[104–106] In the case of tRNA(Leu5), the Rho-independent terminator of its primary transcript *leuX* is removed by PNPase.[107] Moreover, elimination of defective tRNA precursors and tRNA degradation are other two important tasks of PNPase.[20]

PNPase, along with RNase R, is further essential for rRNA quality control, being responsible for the elimination of RNA molecules that appear after the incorrect maturation of 16S and 23S rRNAs.[16] Additionally, PNPase is implicated in 3′ end maturation of 16S rRNA.[108] Recently, it was verified that PNPase can also process the 3′ end of purified pre-small subunits of 16S rRNA *in vitro*.[109] During stationary phase, in *Deinococcus radiodurans*, rRNA decay is performed by RYPER, a ribonucleic complex formed by PNPase, the RNA-binding protein Rsr (a Ro autoantigen ortholog) and the small noncoding Y RNA.[110,111] Y RNA acts as the scaffold and blocks the KH/S1 domains of PNPase, reducing its ability to interact with ssRNA, thus making the more structured rRNAs the newly preferable targets. Interestingly, a similar complex associating PNPase,

Rsr and a ncRNA was also found in *Salmonella* Typhimurium, an evolutionary distant species of *D. radiodurans*. However, its RNA degradation properties are still elusive.[111] Putative Y RNAs also were identified in more than 250 bacteria and phages encoding a Ro ortholog,[112] likely indicating their potential to build RYPER-like complexes for rRNA degradation.

PNPase seems to be particularly relevant for stability control of non-coding RNAs across many distinct species. PNPase-dependent degradation of sRNAs was initially detected in *S.* Typhimurium and *E. coli*.[17,113] The chromosomal encoded MicA and RybB sRNAs that regulate outer membrane proteins are destabilized by PNPase, particularly in the stationary phase.[17] In strong contrast, RNase II and RNase R were not shown to play any role on this regulation. It also was shown that PNPase-mediated activity against sRNA does not necessarily require association within the degradosome. In addition, in the absence of its target *ompA*, the small RNA MicA is still degraded by PNPase.[17] In particular, it was found that sRNAs free of Hfq binding are preferably degraded by PNPase.[19,31] In bacteria, the RNA chaperone Hfq stabilizes sRNAs by protecting them from the attack of ribonucleases.[114] Surprisingly, inactivation of Hfq and PNPase results in the strong increase of MicA, GlmY, RyhB and SgrS sRNAs levels.[19] This high affinity for non-coding RNAs is conserved and the hPNPase was also found to selectively and preferentially degrade a microRNA in human melanoma cells.[115] PNPase is also responsible for the maturation of the CRISPR non-coding RNA RliB in *Listeria monocytogenes*.[116] Without PNPase activity, RliB CRISPR is not correctly processed and is not functional. Therefore, PNPase contributes to the DNA interference activity of RliB CRISPR. This work highlighted a new and unexpected function for PNPase in the processing of CRISPR non-coding RNAs. It has also been demonstrated that PNPase can also stabilize some sRNAs. A *pnp* mutant in *E. coli* presented destabilized sRNAs CyaR, RyhB and MicA.[18,117] It was proposed that PNPase probably protects Hfq-bound sRNAs by limiting the access of RNase E to them. PNPase was shown to protect sRNAs from degradation in the presence of Hfq.[19,118] *In vitro* studies showed that this enzyme degrades some sRNAs in the absence of Hfq but binds and is unable to degrade those sRNAs in its presence, forming a tertiary complex with Hfq and sRNAs.[118]

2.4.3 PNPase in Eukaryotes

PNPase is evolutionarily conserved in bacteria, flies, plants and mammals, including humans.[25] The human PNPase/PNPT1 (previously hPNPase[old-35]) also presents 3′–5′ exoribonucleolytic and polymeric activities. Eukaryotic

PNPase contains extra signal peptides at the N-terminus that target the enzyme to mitochondria in mammals and both the mitochondria and chloroplasts in plants.[119] In *Arabidopsis thaliana* chloroplasts, ribonuclease II/R homologue (RNR1) and cpPNPase cooperate in the RNA degradation pathways and double mutants exhibit embryo lethality. Both enzymes are responsible for the two-step maturation of the 3′ ends of mRNA, with RNR1 polishing the 3′ ends created by cpPNPase. cpPNPase is also involved in tRNA 5′ end leader sequence degradation, according to RNA-seq data. Along with human mitochondrial helicase SUV3, hPNPase is able to degrade dsRNA.[120–122] hPNPase also assembles in a complex with SUV3 to degrade mitochondrial mRNA.[123] Additionally, it was found that hPNPase–SUV3 complex is able to degrade polyadenylated structurally abnormal tRNAs, an important pathway in mitochondrial diseases related with tRNA mutations.[124]

Human PNPase also plays many functions in the cytoplasm. PNPT1 (previously known as hPNPase[old-35]) localizes to the mitochondrial intermembrane space where it is involved in the import of chromosomally encoded transcripts into the mitochondrial matrix.[53,125] Recently, it was reported that PNPT1 is released from mitochondria upon mitochondrial outer membrane permeabilization (MOMP), a process that triggers apoptosis.[126] In the cytosol, PNPT1 initiates apoptotic RNA decay through 3′–5′ degradation of unstructured coding and non-coding polyadenylated RNAs. As with bacterial PNPase, PNPT1 degradative activity is blocked by the presence of stable secondary structures in the RNA. Inactivation of mammalian PNPT1 strongly reduces apoptosis as it leads to inhibition of mRNA decay during cell death. *PNPT1* gene expression is also upregulated during terminal differentiation and cellular senescence.[93,127] This RNA-degrading enzyme is ubiquitously expressed in normal tissues and is a type I interferon inducible early response gene able to degrade *c-myc* mRNA and a specific group of miRNAs, as well as to regulate cell cycle arrest and apoptosis.[115,128–130] It is usually associated with growth inhibition and inflammation during aging.[131] Furthermore, as stated above, hPNPase discriminates oxoguanine-containing RNAs from normal RNAs and it is therefore involved in oxidative stress protection in HeLa cells.[132] In 2012, Lin and colleagues reported the first crystal structure of the hPNPase.[133] The structure presented a typical hexameric RNase PH ring capped with a novel trimeric KH pore, contrarily to the S1 pore for nucleic acids binding present in human exosomes. Moreover, there is an evolutionary conserved GXXG motif in the KH domain involved in RNA binding, lacking in exosomes.

The KH domain is also responsible for the regulation of the cleavage channel size, RNA binding and cleaving activities. Accordingly, hPNPase presents a very constricted channel, which corresponds to more efficient RNA binding and cleavage.

2.4.4 PNPase in Pathogenesis and Disease

PNPase has been associated to virulence and pathogenicity functions, even though its role is still not completely understood and may differ among different bacteria. PNPase can control the abundance of virulence related transcripts. For example, in *Streptococcus pyogenes*, a *pnp* mutant presented a decreased decay rate of *sagA* and *sda* transcripts, which means these transcripts are more stable in the absence of the enzyme.[134] Hence, differential mRNA stability is important for virulence regulation, since *sagA* and *sda* are important virulence factors that encode for Streptolysin S and Streptodornase, respectively.

PNPase is also known to affect the bacterial secretion systems, which could account for the impact of PNPase on pathogenesis. PNPase modulates the secretion of Yop effectors by type III secretion system (T3SS) in *Yersinia pseudotuberculosis*.[135] It downregulates the T3SS transcripts and proteins expression levels and controls the secretion rate through a mechanism that requires the RNA-binding surface of its S1 domain.[135,136] This is probably the reason why the *pnp* deletion results in a less virulent strain in cell culture and mouse model infections.[135,136] Likewise, in *Pseudomonas aeruginosa*, PNPase lacking KH and S1 domains affected the T3SS and cytotoxicity. The mutant strain exhibited reduced levels of expression of T3SS components and pilus biosynthesis genes, while T6SS genes were upregulated.[137] In the case of T3SS, PNPase is required for *exsA* translation, a protein that binds T3SS promoter genes and allows the expression of effector proteins, like ExoS. Increased levels of RsmY and RsmZ seem to be responsible for the altered expression of pili and T6SS genes, since ΔKH–S1 mutant presented increased stability of sRNAs RsmY/Z. The authors showed that PNPase binds directly to these sRNAs and this interaction leads to a twitching phenotype and downregulation of T6SS genes.[137] However, *pnp* deletion in *E. coli* O157:H7 led to enhanced transcription of T3SS components; nevertheless, this resulted in a decrease in the adhesion levels to intestinal epithelial cells.[138] In contrast, *pnp* deletion mutant in *Neisseria meningitidis* showed increased bacterial aggregation and adhesion to epithelial cells but caused less bacteremia in a mouse model infection when compared to the wild-type.[139] PNPase was also found to affect *Campylobacter jejuni* adhesion ability.[140]

In *S.* Typhimurium, a point mutation in *pnp* gene resulted in a truncated form of PNPase that showed a greater ability to invade epithelial cells and to establish persistent infections in mice when comparing to the wild-type.[141] The expression of genes from *Salmonella* pathogenicity islands 1 and 2 was also reduced by PNPase mutation, possibly due to the low frequency of stable stem-loop structures that would be preferentially degraded by PNPase.[141] On the other hand, deletion of the gene coding for PNPase led to a decreased gastrointestinal colonization fitness in pigs by *S.* Typhimurium.[142] As previously mentioned, PNPase is very important in the regulation of sRNAs, namely the ones that control the expression of outer membrane proteins and the adaptation to environmental changes.[17] Alterations in the composition of the proteins may lead to increased sensitivity of *S.* Typhimurium lacking PNPase to the harsh gastrointestinal environment in the colonized pigs.[142]

As previously shown, PNPase is a major factor affecting biofilm formation. In the case of *E. coli* C, PNPase is a negative post-transcriptional regulator of the *pgaABCD* operon (responsible for exopolysaccharide production); thus, there is a reduced biofilm formation in cells expressing PNPase.[143] Conversely, another study in *E. coli* K12 showed that a *pnp* mutant is severely impaired in biofilm formation.[30] Unexpectedly, this *pnp* mutant showed an upregulation of antigen-43 transcript, whose corresponding protein is known to promote cell aggregation and impair motility and, as a consequence, promote biofilm formation.[30] Since biofilm formation is a complex process, there might be a plethora of factors that affect it in the absence of PNPase, and more studies need to be performed. In summary, PNPase has pleiotropic effects on the virulence of many bacteria and the molecular mechanisms by which PNPase controls virulence are also variable.

It was found that hPNPase is located in the mitochondrial intermembrane space.[125,132,144] Surprisingly, it was identified as a novel component in the regulation of the import of nuclear-encoded RNAs into the mitochondrial matrix,[53] acting as an RNA receptor to facilitate the translocation between the intermembrane space and the matrix.[145] hPNPase imports RNAs with folded secondary structures, such as RNase P, 5S rRNA and mitochondrial RNA processing RNAs, from the cytosol to mitochondria.[53,125] These RNAs either have short or no 3′ overhangs, thus they are too short for being pulled to the cleavage channel and they are not processed by hPNPase. However, it remains elusive how hPNPase binds to these structural RNAs.[133] This new function could be employed to tackle mitochondrial genetic disorders, as hPNPase could be used to translocate several RNAs to the mitochondria by using a target sequence that interacts with the enzyme.[146] Mutations in *PNPT1* gene (*hPNPase$^{old\text{-}35}$*), which encodes for hPNPase, are present in patients with neurodevelopmental diseases that also have mitochondrial dysfunction.[147,148] Targeting the human PNPase is a good therapeutic strategy against several pathophysiological conditions, namely tumorigenesis,[130] to selectively downregulate RNA expression[115] or to enhance the delivery of anti-miRNA therapeutics to mitochondria.[149] For instance, the overexpression of hPNPase could be a good way to fight melanomas that overexpress c-*myc* mRNA[130] or miRNA-221 and are resistant to interferon.[115]

3. RNase II

RNase II is the prototype of the widespread RNB family of enzymes. Most interestingly, it is common to find different RNase II paralogues in a same organism.[6] As examples, two RNase II-like enzymes (RNase II and RNase R) are found in *E. coli* and three (hDis3, hDis3L and hDis3L2) can be found in human cells (Figs. 1 and 2). The different RNase II homologues, although similar in domain organization, may exhibit different substrate specificities, expression profiles or even function in different cell compartments. Nevertheless, all RNase II-like enzymes display high activity against ssRNA. RNase II is responsible for most of the hydrolytic activity on polyadenylated RNA in *E. coli* crude extracts and RNase II-like enzymes are responsible for the catalytic activity in archaeal and eukaryotic exosomes. However, unlike the other major 3′–5′ exoribonucleases RNase R and PNPase, RNase II is stalled by fairly stable RNA secondary structures. Strikingly, this trait contributes to an apparent paradox: RNase II can in fact protect several RNAs from degradation.[150] This happens because RNase II efficiently removes any single-stranded region downstream an RNA stem-loop, impairing then the accessibility of other exoribonucleases to those substrates. Undoubtedly, RNase II is an important player in RNA metabolism, affecting not only RNA stability but also RNA maturation. Consequently, RNase II-like enzymes are involved in a multitude of regulatory pathways including those controlling several diseases.

3.1 RNase II Function and Regulation

3.1.1 RNase II Activity on RNA

RNase II is a 3′–5′ exoribonuclease that degrades mRNA through a hydrolytic mechanism, in a processive manner yielding 5′-nucleoside monophosphates. It can disrupt short double-duplexes, but its activity is generally blocked by stem-loop structures.[151] The preferable substrate is the homopolymer poly(A) and the target-transcripts must be longer than 10–15 nucleotides, otherwise RNase II degrades mRNA distributively.[12] The end product released has 4 nts, which are posteriorly destroyed by oligoribonuclease,[3] which is a highly active enzyme in the degradation of single-stranded RNA molecules, being responsible for 90% of the hydrolytic activity on poly(A) RNA observed in cell extracts.[152] Strikingly, a major role for RNase II in cells seems to be the protection of mRNA.[150,153] The single-stranded stretches present at the 3′ end of RNAs are rapidly degraded by RNase II due to its high affinity to these substrates; as consequence, it impairs the access of other exoribonucleases since there is no accessible 3′ single-stranded RNA sequence to which they could bind to initiate degradation.[97]

The RNase II family of enzymes is composed of a large number of homologue 3′–5′ exoribonucleases found across all domains of life.[6] Organisms may contain one or more members of the family, which include the close related bacterial RNase II and RNase R and eukaryotic Rrp44/Dis3 proteins. Due to the high homology between RNase II and RNase R, eukaryotic members of the same family are generally referred to as RNase II-like or RNase II/RNase R homologues. The Rrp44/Dis3 is a yeast RNase II enzyme. It is a 110 kDa protein, that shares 19% homology in its sequence with the *E. coli* RNase II and is capable of degrading single-stranded RNA substrates endo- and exonucleolytically in a processive way.[154] On the other hand, humans have three RNase II-like enzyme paralogues: hDis3, hDis3L1 and hDis3L2. hDis3 and hDis3L1 (also known as hDis3L) have different cellular localizations: hDis3, which is enriched in the nucleus and hDis3L, which is exclusively cytoplasmic.[155–157] These two proteins possess an additional PIN domain in the N-terminus, a feature that is absent in the third eukaryotic homologue Dis3L2.[157]

3.1.2 Regulation of RNase II Activity

Acetylation was found to modify RNase II activity. The Lys501 residue of RNase II can be acetylated, leading to a decrease in substrate binding of

RNase II. This residue is close to the catalytic center and its acetylation partially blocks the RNA channel affecting the enzyme activity.[158] Curiously, the acetylation status of RNase II decreases its enzymatic activity but not its stability. Acetylation of Lys501 is a reversible modification that is post-translationally controlled by Pka and CobB, an acetyltransferase and a deacetylase, respectively. High levels of RNase II acetylation are observed during slow growing cells. RNase R is also regulated by acetylation, but only at the stability and not at the activity level.[159] However, it is not known if Archaea and Eukarya members could be under a similar regulation.

3.1.3 Regulation of RNase II Expression

RNase II is a monomer of 72 kDa and 644 amino acids (aa), encoded by the *rnb* gene from *E. coli*. The *rnb* gene is transcribed by two promotors, P1 and P2, and terminates in a Rho-independent terminator 10 nts downstream of the *rnb* stop codon.[160] Unless PNPase is absent, RNase II is not an essential enzyme in *E. coli*,[61] but no other RNase can substitute for RNase II to maintain cell survival during nutritional starvation.[161] RNase II expression is controlled at the transcriptional and post-transcriptional levels. PNPase cleaves and degrades *rnb* mRNA, modulating its expression.[61] The endonucleases RNase III and RNase E perform an indirect control of RNase II levels. In a mutant strain for RNase III, the *pnp* mRNA is not properly cleaved, affecting the levels of PNPase and, therefore, the levels of RNase II. In a strain mutant for RNase E, there is an increase in both levels of *rnb* mRNA and RNase II itself.[162] Additionally, it was found that the protein Gmr, which possesses a PAS domain that functions as an environmental sensor, controls the stability of RNase II. It was observed that RNase II is twice more stable in the absence of *gmr* when compared to the wild-type.[163] However, the molecular mechanism underlying this regulation remains mostly elusive.

3.1.4 RNase II Complexes

In eukaryotes, a multiprotein complex termed exosome is composed of 9–11 protein subunits that cleave RNA molecules, one nucleotide at a time, in a 3′ to 5′ direction.[164–166] Note that the term exosome is also used to describe extracellular membrane vesicles that deliver RNA and protein and enable cell-to-cell communication and signaling.[167] Throughout this review, we are specifically referring to the exosome complex involved in RNA metabolism. This 3′–5′ exoribonuclease complex is responsible mainly for mRNA decay but also in the processing of small RNAs like snoRNA, snRNAs, and rRNAs. The human exosome core is composed of six proteins (hRrp41, hRrp42, hRrp43, hRrp45, hRrp46 and hMtr3) that form a hexameric ring similar in structure to PNPase. Above this hexameric ring are three proteins that contain KH and S1 RNA-binding domains (hRrp4, hRrp40 and Csl4). Remarkably, all these enzymes are catalytically inactive.[165] The catalytic active subunit is in fact an RNase II-like enzyme, either Dis3 or Dis3L1 (depending if it is the nuclear exosome or the cytoplasmic exosome).

As stated above, bacteria can have a multiprotein complex involved in RNA degradation, the degradosome. PNPase is usually the exoribonuclease found in such complex. Similar to what was observed in eukaryotic exosomes, in which an RNase II-like enzyme was responsible for the catalytic activity, it was suggested that bacterial RNase II might associate with the RNA degradosome.[168] However, this interaction only occurs after binding of PNPase and RhlB to the C-terminal domain of RNase E, which then results on a conformational change that allows RNase II to access the degradosome scaffold. Despite this finding, it remains to be elucidated the impact of such association. It is still not clear if their activities in the degradosome are coordinated or exclusive.[168]

3.2 RNase II Structure

The elemental domain organization composed by RNA-binding domains (two cold shock domains (CDS) in the N-terminus and S1 domain in the C-terminus) flanking the RNB catalytic domain can be found in all members of the RNase II family.[12,169] However, additional domains may exist that confer different biochemical properties to its members (Fig. 3). For example, unlike *E. coli* RNase II, the eukaryotic Rrp44/Dis3 homologue contains a PIN domain in the N-terminus that has endonuclease activity.[170–172] Even though bacterial RNase II and RNase R share a high sequence homology, the presence of additional domains specifically found in RNase R, confer to this enzyme the intrinsic capability to degrade structured RNAs that RNase II lacks (see Section 4).

The catalytic region of all RNase II-like enzymes is the RNB domain and it can stretch of about 400 aa in the central region and adopts a typical αβ-fold.[12] Analysis of the crystal structure of *E. coli* RNase II revealed that nucleotides 1–5 of an RNA fragment interact with the anchor region, formed by CSD1, CSD2 and S1 domains, and the final 9–13 nts are clamped in the RNB domain between residues Phe358 and Tyr253.[12] This explains why transcripts must be longer than 10 nts to be processively degraded, being the Tyr253 residue the one thought to be responsible for setting the size of RNase II end product. The substitution of this residue alters the size of the end product from 4 to 10 nts, causing loosening of RNA substrate at the catalytic site.[173]

The structure of RNase II revealed that RNA degradation involves four highly conserved aspartate residues (Asp201, Asp207, Asp209 and Asp210) that coordinate the recruitment of two Mg^{2+} ions for catalysis. Substitution of the conserved residue Glu542 by an Ala, thought to help the removal of the exiting nucleotide upon phosphodiester cleavage, gave rise to a "super-enzyme" with extraordinary catalysis and binding activities. This conformational modification in the RNB domain and reorganization of the RNA-binding interface displayed 100-fold increase in exonucleolytic activity and 20-fold increase in RNA-binding affinity.[174] Later, it was suggested that RNase II may bind to the cytoplasmic membrane via an amino-terminal amphipathic α-helix (NTH) motif that functions as a membrane binding anchor.[175]

Studies on the radiation resistant bacterium *D. radiodurans* identified that the DrR63 protein is an RNase II-like enzyme. Although it possesses an RNB catalytic domain and C-terminal S1 domain like RNase II, it lacks the two CSD in the N-terminal region. Instead DrR63 possesses a helix-turn-helix (HTH) motif. This HTH motif interacts with an extension of C-terminal S1 domain and it was observed that this makes DrR63 able to approach closer to an RNA duplex than *E. coli* RNase II. The length of the RNA-binding path is similar to the observed for RNase II, however the latter has a narrow clamp whereas DrR63 displays an open architecture and a truncated N-terminus.[176]

3.3 RNase II Substrates

3.3.1 RNase II in the Metabolism of Coding RNA

An RNA-seq analysis comparing the wild-type strain with a deletion mutant of RNase II, identified 187 transcripts differentially expressed. Most of the transcripts were downregulated (67%) as opposed to the upregulated (29%).[30] This is probably related to the protection that RNase II can exert on RNA molecules by efficiently removing the 3′ end linear sequence that is required for binding of the other major exoribonucleases.[150] Interestingly, most transcripts affected by RNase II are related to motility and flagellum assembly. The transcript with a higher fold in the Δ*rnb* mutant was antigen-43 or *flu*, known to promote aggregation and inhibit bacterial motility. Surprisingly, RNase II shares more transcripts with PNPase than with RNase R.[30] Additionally, RNase II is involved in the mRNA degradation of stalled mRNAs that lack a stop codon. This process involves the tmRNA-SmpB complex that provides a stop codon to a stalled ribosome in order to recycle the ribosome, in a mechanism that is called *trans*-translation.[177] In *E. coli*, prolonged translational arrest allows mRNA degradation into the A-site of stalled ribosomes. The enzyme that cleaves the A-site codon is not known, but its activity requires RNase II to degrade mRNA downstream of the ribosome.[178]

3.3.2 RNase II in the Metabolism of Non-coding RNA

Even though its major role is the degradation of mRNA, RNase II also functions in the processing of tRNA and other stable RNAs. However, the role of RNase II in these reactions seems very limited. A whole-genome transcriptional study revealed that inactivation of RNase II marginally affected stable RNAs, including tRNAs and sRNAs.[30] In fact, RNase II has not generally been implicated in the degradation of regulatory RNAs.[19,31] Endoribonucleases are thought to be involved in generating the mature 5′ termini while maturation of the 3′ end requires the action of exoribonucleases.[179] Any one of a number of exoribonucleases, including RNases T, D, PH, II, R and PNPase can carry out the trimming reaction of tRNAs *in vivo*, although RNases T and PH are the most effective.[180] Hence, in normal conditions, RNase II activity is not critical for tRNA processing. Similar observations can be made for rRNA processing. Multiple exoribonucleases can catalyze the maturation of pre-16S rRNA.[108] However, RNase R and PNPase are the most important for the elimination of aberrant rRNA fragments under normal growth conditions.[16,181] In contrast, under starvation, RNase II and RNase R are the most important exoribonucleases for rRNA fragment removal.[181] RNase II was found to control RNase PH activity under starvation.[161] RNase PH is very unstable during starvation and stationary conditions with levels of the enzyme decreasing as much as 70%. Yet, the mechanism by which RNase II controls and regulates RNase PH remains incompletely understood but is suggested to correlate with RNase II mediated degradation of ribosomes.[161]

4. RNase R

RNase R is unique amongst the 3′–5′ exoribonucleases in its remarkable capability to degrade RNA secondary structures on its own.

Unlike RNase II, RNase R is not blocked by RNA stem-loops and in contrast to PNPase it does not require association with other proteins (e.g., helicases) to overcome such structures. Furthermore, RNase R is a stress regulator induced in different environmental stimuli being particularly important in the cold shock response, which leads to a strong stabilization of RNA structures. Not surprising, RNase R is highly effective against a wide variety of substrates. RNase R is a key enzyme involved in different surveillance mechanisms that work to eliminate unnecessary mRNAs, defective tRNA, aberrant rRNA molecules and damaged ribosomes.[16,20,182] Overall, RNase R arises as an important quality control enzyme for translation. Much work on RNase R elucidated a novel mode of regulation of RNA degradative enzymes. Additionally, RNase R is now used as a routinely tool in molecular biology, a feature particularly important for the study of circular RNAs (circRNAs). The continuous study of this unique enzyme will certainly continue to expand the knowledge of RNA-dependent pathways.

4.1 RNase R Function and Regulation

4.1.1 RNase R Activity on RNA

RNase R is a hydrolytic exoribonuclease belonging to the RNase II family of proteins. RNase R is a highly processive and sequence-independent enzyme that degrades RNA releasing nucleoside monophosphates, having a dinucleotide as the end product of digestion.[183] Although it shares the family typical RNB catalytic domain, RNase R comprises distinctive features from RNase II. For instance, its ability to degrade structured RNA substrates making use of an intrinsic helicase activity requiring only a single-stranded 3′ overhang of at least 7 nts for effective binding.[184] Substrates with 3′ overhangs as shorter as 3 nts can also be degraded, although not as efficiently.[185] RNase R is involved in several quality control mechanisms that act to remove defective RNAs. This exoribonuclease is effective against a wide range of substrates, from highly structured mRNAs (as those containing REP sequences) to highly structured ribosomal RNAs. RNase R homologues are widespread in nature and ubiquitously important in various organisms.[3] Comprising both helicase and exoribonuclease activities make it a unique enzyme in the RNA degradation landscape. In recent years such outstanding features have led for RNase R to be utilized as a molecular biology tool, specifically in the study of eukaryotic circular RNAs. The 5′ and 3′ ends of circRNAs are not free, instead they are covalently bound together, hence yielding a circular RNA molecule.[186] The expression of circRNAs is generally developmental stage, cell-type and tissue specific, that arouse great interest on the research community. Since circRNAs have no unprotected extremities they are essentially exoribonuclease-resistant. RNase R was found to be the best exoribonuclease for the methodical digestion of eukaryotic linear RNAs providing an enrichment in circRNAs for further downstream experiments.[187,188]

4.1.2 Regulation of RNase R Stability

The *rnr* gene that encodes RNase R, is part of an operon containing: *nsrR* (transcriptional regulator), *rnr* (exoribonuclease), *rlmB* (rRNA methyltransferase) and *yjfI* (unknown function).[14] Operon transcription is thought to be driven by a σ^{70} promoter and transcript processing and decay is mainly regulated by RNase E.[189] RNase R levels change in response to different physiological conditions, namely there is a 3- to 10-fold increase during various stresses such as cold shock or stationary phase of growth.[21,190,191] However, overexpression of RNase R becomes hazardous to the bacterial cell, possibly because of its ability to degrade virtually all cellular RNAs.[183] Hence, it was hypothesized that a tightly regulated mechanism for controlling RNase R levels was in place. It was firstly discovered that such regulation was accomplished by variations in RNase R stability and that the stabilization correlated with increasing protein levels in response to stresses.[21,190–192]

Strikingly, the variations in RNase R stability were found to be controlled by the post-translational acetylation status of a single lysine residue (Lys544). RNase R becomes highly unstable when Lys544 is acetylated and is degraded; in contrast, when Lys544 is not modified RNase R is very stable.[159] This nicely correlates with the fact that RNase R from exponential growing cells was shown to be acetylated, whereas stationary-phase RNase R is not. This acetylation is dependent on the product of the *yfiQ* gene which encodes lysine acetyltransferase Pka. In fast growing cells, Pka carries out Lys544 acetylation destabilizing RNase R. However, during stationary phase or cold shock conditions Pka is absent and acetylation activity decreases dramatically.[193]

Acetylation of the Lys544 residue stimulates the binding of tmRNA–SmpB which in turn promotes proteolysis through the recruitment of Lon and HslUV proteases to the N-terminal region of RNase R.[194] Conversely, tmRNA–SmpB was shown to bind only weakly to RNase R during stationary phase of growth, when Lys544 is not acetylated. This leads to a protein stabilization and consequently increases RNase R levels, allowing its crucial action during these stress conditions. RNase R was further found to be highly associated with ribosomes during exponential growth.[195] RNase R binding to ribosomes fully stabilizes the protein whereas its free form turns over extremely fast (~2 min). Intriguingly, tmRNA–SmpB is required for both RNase R association to ribosomes and turnover of the free form of the enzyme, although the two processes are independent. This allows the free form of RNase R to carry out its well-known functions as a stress response post-transcriptional regulator by acting on specific RNA substrates.

4.1.3 RNase R Complexes

As RNase R has intrinsic helicase activity, there is no specific requirement for association with RNA helicases or other proteins in multiprotein complexes to succeed in the degradation of double-stranded RNAs. In fact, RNase R is not found in "normal" degradosomes. However, RNase R replaces PNPase as the exoribonuclease present in "specialized" degradosomes, formed under stress conditions. RNase R interacts with the endoribonuclease RNase E in the degradosome of the cold-adapted bacterium *Pseudomonas syringae* Lz4W.[196] Curiously, RNase R is also found in the stationary-phase degradosome in *E. coli*.[197] We could speculate that other conditions that result in the induction of RNase R expression may lead to its incorporation in similar multiprotein complexes. In addition, RNase R was reported to interact with the 30S subunit, specifically with the ribosomal protein S12.[198] However, the importance of this contact remains mostly unknown to date.

4.2 RNase R Structure

RNase R is a 92 kDa multidomain protein with a modular arrangement like that of the RNase II family members. The catalytic RNB domain is flanked by the RNA-binding domains, the cold shock domains CSD1 and CSD2 at the N-terminus and the S1 domain at the C-terminus (Fig. 3).[173] Strikingly, RNase R is bigger than RNase II since it comprises at least two additional domains: a K/R rich domain and a helix-turn-helix domain (HTH) at the C- and N-terminal regions, respectively.[199,200] Mutational studies identified key residues for the nuclease activity located in the active center of the RNB domain. Of notice, a Asp280Asn mutant (similar to the Asp209Asn mutant of RNase II previously reported) exhibited no exoribonucleolytic activity.[12] Furthermore, the Tyr324 residue was identified as essential for keeping a dinucleotide as the final product of digestion.[201] A distinctive feature of RNase R is the much stronger binding of the catalytic channel for the

RNA substrates in comparison to RNase II, consequently providing a stronger pull on the RNA molecule.[202] Strikingly, a mutant RNase R expressing only the RNB catalytic domain can degrade a fully complementary double-stranded RNA molecule, whilst the presence of the RNA-binding domains restores the requirement for a linear 3′ overhang. Therefore, it was suggested that the RNA-binding domains (CSD1, CSD2 and S1) would function as discriminatory domains to select RNA molecules with 3′ extensions as RNase R substrates.[201] Moreover, the flanking RNA-binding domains are responsible for the intrinsic helicase activity of RNase R.

Two additional regions were meanwhile identified in RNase R: a Walker A motif within the C-terminal K/R rich patch and a Walker B motif within the N-terminal CSD2 domain. Both the Walker A and B motifs are responsible for forming an ATP binding site when near each other. Interestingly, since both motifs are in opposite extremities of the protein, this means that the N- and C-terminal regions must come together. A conformational change is thought to occur upon dsRNA binding, bringing the Walker A and B motifs together thus forming the ATP pocket. ATP binding, but not its hydrolysis, then stabilizes the conformational rearrangement and triggers the helicase activity, thus stimulating RNA strand binding to CSD and S1 domains.[199]

Crystallization of full-length RNase R proved to be a challenging task. Nonetheless, a truncated form lacking the N-terminal helix-turn-helix domain and the C-terminal K/R rich region (RNase R ΔHTH-K mutant) yielded crystals that allowed the resolution of its structure.[200] RNase R ΔHTH-K variant exhibited only a slightly slower *in vitro* activity towards a dsRNA with a 10 nucleotide 3′ overhang when compared to the full-length protein, thus confirming that truncation did not significantly affect the overall enzymatic activity.[200–202] The crystallized protein also bore a spontaneous mutation (Ala131Val) that enhanced crystallization, but why this was the case remains unclear since the residue is in a disordered loop. The three-dimensional structure of other members of the RNase II family had already been determined for Rrp44, Dis3L2 and RNase II itself, which served as basis for comparative studies.[12,166,203] Superimposition of the RNB domain of RNase R with those of RNase II and Rrp44 revealed that they share an overall similar structure. Strikingly, the auxiliary RNA-binding domains adopted a different position and orientation, suggesting either that these domains are flexible or, perhaps more likely, that the RNases have different RNA-binding modes.

Insights into the hydrolysis mechanism confirmed it is identical to that of RNase II.[12] Specifically, a two-metal ion catalytic mechanism where one Mg^{2+} is bound by residues Asp272 and Asp281 in the RNB domain, whereas a second Mg^{2+} binds only after RNA binding in the active site and is therefore absent from the 3D structure. Moreover, residues Tyr324 and Tyr383 match two tyrosine residues at the same position of RNase II that stack the 3′ end nucleotides.[12] It is likely that Tyr324 and Tyr383 are, therefore, responsible for setting the end product of the digestion as a dinucleotide, corroborating previous biochemical data.[200–202]

Another important feature of RNase R structure is a tri-helix region located in the RNB domain that serves as a wedge for the unwinding of RNA. Curiously, in the RNA-bound crystal structure reports of the RNase II family members, the RNB tri-helix region was found to interact with the substrate RNA on Dis3L2 and Rrp44,[203,204] however this interaction was not observed for the RNase II structure, which argued for the importance of the tri-helix region for the unwinding activity. In fact, mutational and biochemical experiments revealed that an RNase R deletion mutant lacking the whole tri-helix region is ineffective in unwinding its substrate.[200] The energy for the unwinding of the RNA is therefore proposed to derive from the hydrolysis of the RNA itself, and not from ATP hydrolysis as in other RNA helicases. The hydrolysis of substrate further pulls the 3′ end of the molecule to the active site while the wedge region helps unwind the RNA.[199,200]

Published data showed that both RNase II and Dis3L2 bind a ssRNA molecule through the S1 and CSD1 motifs forming a top channel that caps the RNB catalytic motif, whereas the crystal structure of Rrp44 bound to ssRNA revealed a side channel formed by CSD1 and RNB. Strikingly, the crystal structure of RNase R revealed that it possesses two open channels: one top channel (S1 and CSD1) and one side channel (CSD1 and RNB), which is a unique feature among other members of the RNase II family.[200] Based on the crystallographic data two models for RNase R action can be proposed, depending on which one of the two channels is responsible for threading the RNA to the active site or serve as an exit for the non-sessile 5′ end that is unwound during degradation. One model is based on the Rrp44–RNA structure and contemplates the side channel as the entry channel, whereas a second model based on the RNase II-RNA structure suggests the top channel as the entrance. The latter is more likely to be true since measurements of both RNase R channels indicate that the top channel's width is sufficient for accommodating dsRNA, contrasting with the narrower side channel where only ssRNA could bind. Undoubtedly the resolution of RNase R crystal structure is a milestone in understanding the complex behavior of this remarkable enzyme. Nonetheless, a co-crystallization of RNase R with a structured RNA substrate would certainly allow for a refined model of its action, as happened with other members of the RNase II family of enzymes.

4.3 RNase R Substrates

4.3.1 RNase R in the Metabolism of Coding RNA

Under normal laboratorial conditions in bacteria grown to exponential phase, an RNA transcriptomic study revealed that the single inactivation of RNase R affects only ~6% of *E. coli* open reading frames (ORFs).[30] Nevertheless, RNase R variants with mutations within or close to the entrance of the nuclease channel were shown to suppress the mRNA turnover defects of an RNase E mutant, which raises the question if RNase R could be more important for mRNA decay than previously thought.[205] RNase R has been shown to participate in the elimination of REP-containing mRNAs and in the polyadenylated-dependent degradation of the *rpsO* mRNA.[29,206] RNase R is induced in response to several stresses.[21,190,191] This enzyme seems particularly important in the degradation of mRNAs under these harsh conditions, namely in adaptation of cells to temperature downshift.[68]

An important bacterial strategy to survival is adaptation to temperature fluctuations, which involves a rapid and extensive reprogramming of gene expression. Bacteria exposed to significantly sub-optimal temperatures develop a cold shock response that is characterized by an acclimation phase during which cell growth temporarily stops. During this phase, even though translation is drastically reduced, some cold shock proteins are produced, which helps cells to adapt to low temperature.[207] RNase R is one of these cold shock proteins and its induction occurs through mRNA and protein stabilization and increased translation after the temperature downshift.[159,189] Inactivation of RNase R leads to growth deficiency at low temperatures in different bacteria.[189,208] One of the main consequences of cold shock is the stabilization of RNA secondary structures which correlates with lower translation levels. For a long time, it was suspected that RNase R was important in mRNA decay at low temperature. In *Legionella pneumophila*, inactivation of RNase R results in the accumulation of structured RNAs specifically in cold shock.[209] *E. coli* RNase R, but not RNase II or PNPase, can substitute for the RNA helicase CsdA, an essential gene involved in mRNA decay in low temperature.[210] These two enzymes were suggested to share some common target mRNAs.[211] The role of RNase R in mRNA degradation during acclimation was recently confirmed using global methods to monitor the genome-wide changes in translation.[68] This study demonstrated that RNase R is part of a surveillance system that eliminates the poorly translated structured mRNAs that are stabilized by low temperature and may compete to access ribosomes, thus playing a key role for translation recovery after cold shock.

4.3.2 RNase R in the Metabolism of Non-coding RNA

A wide range of RNA substrates is targeted by RNase R. However, because of its natural ability to degrade RNA secondary structures, RNase R has major roles in the quality control and degradation of rRNA.[16,212] Bacterial ribosomes are extremely stable in exponential growing cells. However, under stresses that lead to slow growth, there is a lower demand for functional ribosomes because of a global decrease in translation. Therefore, there is a need for recycling the constituents of the macromolecules to ensure energy efficiency and cell survival. Endoribonucleases break down the large rRNA molecules internally to smaller fragments that are subsequently degraded by 3′–5′ exoribonucleases. It has been shown that elimination of intermediate rRNA fragments during starvation conditions is carried out by RNase R and RNase II. Accumulation of these fragments is hazardous to the cell since there is no recycling of ribonucleotides.[212,213]

As mentioned above, an RNase R and PNPase double mutant is synthetic lethal. This inviable phenotype was shown to arise shortly after 16S and 23S rRNA fragments accumulated to high levels in a conditional RNase R^- $PNPase^-$ double mutant.[16] These fragments were also suggested to be generated after initial endonucleolytic cleavages. Both RNase R and PNPase were thus identified as being part of the same quality control mechanism that eliminates aberrant rRNAs during steady-state growth as soon as they are generated. Failing to do so interferes with correct ribosome maturation leading to cell death. Additionally, RNase R was described to take part in a novel late-stage quality control mechanism that also involves YbeY, a newly identified endoribonuclease.[182] Depletion of YbeY leads to accumulation of misprocessed 16S rRNA precursor and immature 30S subunits, which argues for a role in the processing of rRNA. Strikingly, YbeY and RNase R were found to mediate the specific *in vitro* degradation of 70S ribosomes baring defective 30S subunits. The immature small 30S subunits were found to be the trigger that allowed YbeY to distinguish between defective and non-defective ribosomes. The latter are thus thought to be initially targeted for endonucleolytic cleavages by YbeY followed by RNase R unwinding and processive digestion.[182] The requirement for both YbeY/RNase R for this quality control mechanism is substantiated by the observation that *ybeY* and *rnr* genes were shown to have a strong genetic interaction, which could also explain the high conservation of both proteins throughout evolution.[214]

The role of RNase R in rRNA and ribosome homeostasis is further underlined by the fact that it participates in the 3′ end maturation of the 17S rRNA (16S rRNA precursor), a functional overlap with PNPase, RNase II and RNase PH.[108] This finding places RNase R not only as part of degradation and surveillance machinery, but also on the list of 16S rRNA processing enzymes.[109] A role in the maturation of the 3′ end of rRNAs had been previously described for RNase R homologues in the Gram-negative bacteria *P. syringae* during cold shock and in the eukaryotic organism *A. thaliana*, which argues in favor of a conserved widespread function.[196,215]

RNase R also participates in tRNA quality control by degrading defective tRNAs in *E. coli*.[20,184] Addition of either poly(A) or CCACCA tails to the tRNA stimulates the RNase R activity against tRNA.[216] This exoribonuclease seems to have a more significant role as a degradative enzyme rather than processing enzyme of tRNA in *E. coli*. However, RNase R was suggested to play an important function in tRNA processing in other bacteria. Shorter $tRNA^{Cys}$ species accumulated only when RNase R was also absent, suggesting that this RNase R has a principal role in the quality control of tRNAs in *B. subtilis*.[217] RNase R was also shown to be the responsible enzyme to remove tRNA 3′-trailers and generate mature 3′ ends in tRNAs from *Mycoplasma genitalium*.[218]

RNase R is also required for the correct processing of the tmRNA in *E. coli* during cold shock and in the degradation of the tmRNA in *C. crescentus* during cell cycle.[190,219] This RNA molecule holds the unique bifunctional ability of acting as both a tRNA and mRNA, and in concert with the protein SmpB directs the stalled ribosome to its own ORF tagging the aberrant protein for proteolysis. Moreover, RNase R was shown to co-purify with the tmRNA–SmpB complex.[220,221] RNase R is suggested to degrade the aberrant nonstop mRNA and thus play a key role in the *trans*-translation mechanism of protein quality control.[222,223] Surprisingly, RNase R is not found to greatly affect the stability of small RNAs.[17,19] Nevertheless, it was reported that RNase R is involved in the degradation of both the sRNA SR4 and its target *bsr*G mRNA of *B. subtilis*.[224]

We anticipate that the continuous study of sRNA stability pathways will highlight the role of RNase R in the metabolism of many more regulatory RNAs.

4.4 RNase II/RNase R in Eukaryotes

E. coli RNase II (EcRNase II) has a very ancient origin with many organisms sharing homologues with it, like *Drosophila melanogaster* (named DmTAZ), humans (hDis3), yeast *S. cerevisiae* (ScRrp44), *Saccharomyces pombe* (SpDis3), *Caenorhabditis elegans* (CeDis-3), and *A. thaliana* Rrp44.[225] RNA degradation in eukaryotes is more complex compared to that in prokaryotes, given the different complexity of both types of cells. The eukaryotic cell is compartmentalized into two main regions, the nucleus and the cytoplasm. Compartmentalization can change the RNAs fate, and RNA degradation is important in both places.[226]

Yeast Rrp44/Dis3 has a high degree of conservation between yeast and humans and with an indispensable role in the function of the exosome complex.[227] The hDis3 is present mostly in the nucleus of human cells but a minor fraction has been also found in the cytoplasm.[155,156] On the other hand, hDis3L1 is exclusively cytoplasmatic. Each one associates with either exosome complex in each of these cellular compartments through their N-terminal PIN domain and have a broad list of substrates such as mRNAs, snoRNA precursors, introns, long non-coding RNAs, microRNAs and tRNAs.[155–157,228] A third homologue, hDis3L2 is also found in human cells. Dis3L2 is well conserved between bacteria, fission yeast, humans and plants. Dis3L2 contrarily to the other two enzymes lacks the PIN domain region and does not associate with the exosome. Localized in the cytoplasm of the cell, it acts in the degradation of mRNA and various non-coding RNAs and is active against structured RNA.[157,229–231] Uridylation of the RNA was shown to stimulate the Dis3L2-mediated degradation of different RNAs, including abundant non-coding RNA, such as 5S rRNA, tRNA, snRNA, snoRNA.[229,232] Several other targets are described like the pre-let-7 in mouse embryonic stem cells,[233] miR-27a in human cells[157,234] and CG2678 (a transcription factor) and pyrexia (a TRP cation channel) in *D. melanogaster.*[231] It is thus fair to say that Dis3L2 has a general and broad role in RNA metabolism across species and different types of RNA molecules.

The yeast mitochondrial degradosome complex (mtEXO) is a system that allows 3′–5′ decay in the yeast mitochondria.[235] It is a minimal system composed of two critical subunits, a 3′–5′ exoribonuclease from the RNase II family named Dss1 and the nucleotide triphosphate (NTP)-dependent RNA helicase Suv3, important in the unwind of the RNA substrates and involved in general RNA processing, degradation and surveillance.[235] Both genes are nuclear encoded and localize to mitochondria.[102,236] The RNase in this system functions in a progressive way with the dependence of ATP, releasing nucleoside monophosphates and leaving 4 nts, similar to the bacterial RNase II.[151,237] An important aspect of Dss1, making it a special member of the RNase II family, is its intrinsic association with Suv3 helicase. The presence of the helicase makes the Dss1p subunit exoribonucleolytic activity increase approximately 10-fold.[237] So far, no orthologs of Dss1p were reported in higher eukaryotes; hence, human and plant mitochondria lack the Dss1 exoribonuclease.

RNR1 is encoded by the *Arabidopsis* At5g02250 gene.[6,215] RNR1 is a ubiquitous protein being present in the mitochondria where it completes mRNA 3′ maturation initiated by PNPase, but also involved in the processing and maturation of chloroplast rRNAs, specifically 23S, 16S and 5S rRNAs.[121,215] *A. thaliana* AtRRP44A is the functional homologue of yeast Rrp44/Dis3. It is crucial for plant survival and is required for RNA decay.[238] AtRRP44B/SOV showed a different target RNA profile when compared with AtRRP44A and the same exosome core components.[238] It encodes a conserved, cytoplasmic localized Rrp44/Dis3-like protein.[225] Unlike AtRRP44A, it lacks a PIN domain, which suggests that AtRRP44B/SOV works independently from the exosome and more, that it might have similarities with the fission yeast Dis3L2.[238]

4.5 RNase II/RNase R Family Members in Pathogenesis and Disease

RNase II-like enzymes activity have been linked to pathogenesis and disease mechanisms. For example, a Gram-negative bacterium, *Photorhabdus temperata*, an insect pathogen that uses the nematode *Heterorhabditis* as a vector, produces a wide variety of virulence factors and presents hemolytic activity.[239] Through construction of a mutant library, it was found that inactivation of the gene *pte_02637* encoding for RNase II, resulted in a decreased production of extracellular factors, motility and reduced virulence of the bacterium. The stress-inducible RNase R has also been associated with the infection ability of many pathogens. Similar to what happens in *E. coli*, RNase R is also a cold shock protein essential for growth at low temperatures in the pathogens *L. pneumophila*, *Pseudomonas putida* and

P. syringae.[209,240,241] Furthermore, it is also responsible for 16S and 5S rRNA maturation and tmRNA decay in these species.[196,240] In *Aeromonas hydrophila,* RNase R is also a cold shock protein whose absence reduces cell motility and attenuates bacterial virulence in a mouse model.[242] Curiously, RNase R is the only exoribonuclease in *Mycoplasma genitalium* and is therefore essential for normal growth.[243] Inactivation of RNase R in *C. jejuni* reduces this bacterium ability to host cell adhesion and invasion, thus impacting bacterial virulence.[244] In *Shigella flexneri*, RNase R is required for the expression of several invasion factors. Moreover, in both *Shigella* spp. and enteroinvasive *E. coli*, RNase R was shown to be crucial for the production of the virulence related Vir proteins, that are necessary for the correct virulence phenotypes.[245]

Other RNase II-like proteins are also implicated in human diseases. A putative RNase II homologue in *Plasmodium falciparum* (PfRNase II) was implicated in malaria pathogenesis.[246] Malaria parasites have strategies to cope with environmental changes. Those strategies rely on specific families of genes that allow phenotypic variation and on the differential expression of those genes. One of these families of genes is the virulence and pathogenesis-linked *var* gene family that is divided into three major subtypes (*upsA*, *upsB* and *upsC*).[247] Severe malaria is associated with *upsA* and/or *upsB* while asymptomatic malaria is associated with *upsC*.[247] It was found an upregulation of *upsA* genes in *P. falciparum* parasites that had a defective PfRNase II, suggesting a clinical relevance of this exoribonuclease in the malaria.[246,248]

The absence of Dis3/TAZMAN in *S. cerevisiae* and *Drosophila* has been associated with lethality, due to failures in mitosis.[249] Dis3 is also extremely associated with cancer. One of the more critical relations between Dis3 and cancer is in multiple myeloma (MM) where Dis3 is recurrently mutated.[250] There are also reports that state the discovery of Dis3 repeatedly mutated in acute myeloid leukemia (AML), which is another type of blood related cancer.[249,251] Interestingly, the full-length Dis3 is the main transcript differentially expressed in AML patient cells, which highlights the role of Dis3 in this blood cancer progression.[249] Dis3 was also identified as a player in colorectal cancer and in two different types of melanoma, specifically superficial spreading melanoma (SSM) and nodular melanoma (NM) where it presents different expression levels being under expressed in SSM and overexpressed in NM. From this difference it was concluded that the role of Dis3 differs with the context of specific cancers.[249,250] Dis3L2 has also been shown to be involved in several human diseases. The siRNA knockdown of hDis3L2 caused defects in mitosis and abnormal expression of various mitotic proteins and that hDis3L2 accompanied by an increase in cell proliferation.[230,252] These characteristics are consistent with the overgrowth and tumor susceptibility phenotype seen in Perlman syndrome, a rare congenital overgrowth syndrome, proving that Dis3L2 has major cancer implications.[230,252]

5. CONCLUDING REMARKS

The bacterial 3′–5′ exoribonucleases PNPase, RNase II and RNase R are the major degradative enzymes involved in RNA metabolism. These RNA-regulating proteins are conserved and widespread among different phyla. Although the ancestral genes are the same, subsequent mutation has resulted in sequence divergence and domain acquisition or loss. Notwithstanding that the enzymes follow the basic domain organization found in the ancestral gene, homologues show incorporation or loss of some domains. This is particularly evident when analyzing members of the RNase II family (Fig. 3). Furthermore, some organisms may express up to three paralogues of RNase II, which result from gene duplication. RNase II paralogues diverge in their function and/or cell location, as observed in *E. coli* or human cells. On the other hand, no paralogues are found for PNPase (Figs. 1 and 2).

These fascinating enzymes are active against a wide range of substrates, from the more labile coding mRNA to more stable RNAs (tRNA, rRNA and sRNA). Even though PNPase, RNase II and RNase R share substrates or appear to have overlapping roles, they can exhibit different affinities for the same RNA molecule (Fig. 4). This is true even for very close related exoribonucleases like RNase II and RNase R, which share a high sequence and structure homology but display different catalytic activities and have different substrate preference. RNase R is highly effective against structured RNA while RNase II is blocked by RNA secondary structures. A similar observation can be made with eukaryotic RNase II-like enzymes: Dis3 and Dis3L2 are paralogues but only Dis3L2 recognizes uridylated RNAs and mediates their degradation. Bacterial PNPase shows a high affinity to small non-coding RNAs, a feature that is found to be conserved, as human PNPase is key in the degradation of microRNAs. Remarkably, we can verify that the similarities between prokaryotes and eukaryotes are much higher than initially expected.

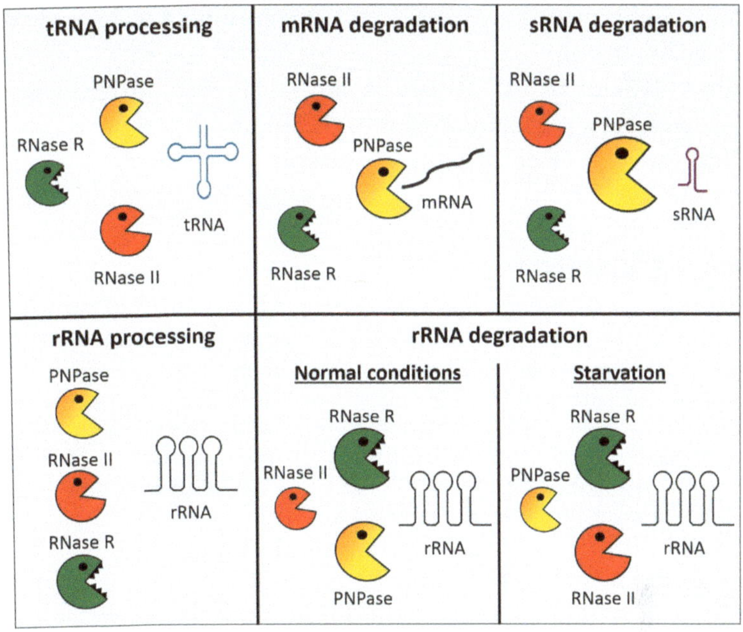

Fig. 4 Affinity of *E. coli* major exoribonucleases to specific substrates. Involvement of PNPase, RNase II and RNase R in degradation or processing of different RNA molecules. The proximity of each exoribonuclease to the substrate represents its affinity towards that specific RNA molecule. The size of each exoribonuclease is related to its importance in the activity against a specific substrate.

The study of PNPase, RNase II, RNase R and related homologues have greatly contributed and continue to expand our knowledge on RNA processing and degradation. The identification of Dis3L2, the most recent RNase II-like enzyme identified, has contributed to unravel novel RNA degradation pathways that were not previously envisaged. The continuous study of these enzymes has elucidated how the fate of many RNAs is controlled, which allows the comprehension of specific genetic pathways. This has been particularly important to understand how 3′–5′ exoribonucleases affect virulence and several disease-related processes, through the identification of previously unrecognized RNA-based pathways. The structures of PNPase, RNase II, RNase R and homologue proteins have been solved at an increasing pace providing crucial details to fully understand their mechanisms of action and the many novel regulatory pathways involving exoribonucleases that have been discovered. In addition, we now have a much better knowledge on the post-translational regulation of bacterial 3′–5′ exoribonucleases. This raises the question if the archaeal and eukaryotic counterparts may have similar mechanisms for controlling their activity and/or stability.

PNPase, RNase II and RNase R are fascinating enzymes that have helped to push forward many studies on RNA and RNA-based pathways. This allows us to foresee that the analysis of 3′–5′ exoribonucleases will remain a dynamic area with many surprises kept for the years to come.

ACKNOWLEDGMENTS

This work was financially supported by Project LISBOA-01-0145-FEDER-007660 (Microbiologia Molecular, Estrutural e Celular) funded by FEDER funds through COMPETE2020—Programa Operacional Competitividade e Internacionalização (POCI) and by national funds through FCT—Fundação para a Ciência e a Tecnologia, including the Program IF (IF/00961/2014) and grants PTDC/IMI-MIC/4463/2014 and PTDC/BIA-MIC/32525/2017 to JMA. RFdS and APQ are recipients of FCT Doctoral fellowships (PD/BD/105733/2014 and PD/BD/135487/2018, respectively) in frame of the ITQB PhD Program in Molecular Biosciences.

REFERENCES

1. Carpousis AJ, Luisi BF, McDowall KJ. Endonucleolytic initiation of mRNA decay in Escherichia coli. *Prog Mol Biol Transl Sci.* 2009;85:91–135. https://doi.org/10.1016/S0079-6603(08)00803-9.
2. Mackie GA. RNase E: at the interface of bacterial RNA processing and decay. *Nat Rev Microbiol.* 2013;11(1):45–57. https://doi.org/10.1038/nrmicro2930.
3. Andrade JM, Pobre V, Silva IJ, Domingues S, Arraiano CM. The role of 3'-5' exoribonucleases in RNA degradation. *Prog Mol Biol Transl Sci.* 2009;85(C):187–229. https://doi.org/10.1016/S0079-6603(08)00805-2.
4. Durand S, Condon C. RNases and helicases in Gram-positive bacteria. *Microbiol Spectr.* 2018;6(2):1–17. https://doi.org/10.1128/microbiolspec.RWR-0003-2017.
5. Mohanty BK, Kushner SR. Enzymes involved in posttranscriptional RNA metabolism in Gram-negative bacteria. *Microbiol Spectr.* 2018;6(2):1–16. https://doi.org/10.1128/microbiolspec.RWR-0011-2017.
6. Zuo Y, Deutscher MP. Exoribonuclease superfamilies: structural analysis and phylogenetic distribution. *Nucleic Acids Res.* 2001;29(5):1017–1026.
7. Kumar S, Stecher G, Li M, Knyaz C, Tamura K. MEGA X: molecular evolutionary genetics analysis across computing platforms. *Mol Biol Evol.* 2018;35(6):1547–1549. https://doi.org/10.1093/molbev/msy096.
8. Szklarczyk D, Morris JH, Cook H, et al. The STRING database in 2017: quality-controlled protein-protein association networks, made broadly accessible. *Nucleic Acids Res.* 2017;45(D1):D362–D368. https://doi.org/10.1093/nar/gkw937.
9. Portnoy V, Evguenieva-Hackenberg E, Klein F, et al. RNA polyadenylation in Archaea: not observed in Haloferax while the exosome polynucleotidylates RNA in Sulfolobus. *EMBO Rep.* 2005;6(12):1188–1193. https://doi.org/10.1038/sj.embor.7400571.
10. Sorenson RS, Deshotel MJ, Johnson K, Adler FR, Sieburth LE. Arabidopsis mRNA decay landscape arises from specialized RNA decay substrates, decapping-mediated feedback, and redundancy. *Proc Natl Acad Sci USA.* 2018;115(7):E1485–E1494. https://doi.org/10.1073/pnas.1712312115.

11. Symmons MF, Jones GH, Luisi BF. A duplicated fold is the structural basis for polynucleotide phosphorylase catalytic activity, processivity, and regulation. *Structure*. 2000;8(11):1215–1226.
12. Frazão C, McVey CE, Amblar M, et al. Unravelling the dynamics of RNA degradation by ribonuclease II and its RNA-bound complex. *Nature*. 2006;443(7107):110–114. https://doi.org/10.1038/nature05080.
13. Donovan WP, Kushner SR. Polynucleotide phosphorylase and ribonuclease II are required for cell viability and mRNA turnover in Escherichia coli K-12. *Proc Natl Acad Sci USA*. 1986;83(1):120–124.
14. Cheng ZF, Zuo Y, Li Z, Rudd KE, Deutscher MP. The vacB gene required for virulence in Shigella flexneri and Escherichia coli encodes the exoribonuclease RNase R. *J Biol Chem*. 1998;273(23):14077–14080.
15. Li Z, Deutscher MP. The role of individual exoribonucleases in processing at the 3' end of Escherichia coli tRNA precursors. *J Biol Chem*. 1994;269(8):6064–6071.
16. Cheng Z-F, Deutscher MP. Quality control of ribosomal RNA mediated by polynucleotide phosphorylase and RNase R. *Proc Natl Acad Sci USA*. 2003;100(11): 6388–6393. https://doi.org/10.1073/pnas.1231041100.
17. Andrade JM, Arraiano CM. PNPase is a key player in the regulation of small RNAs that control the expression of outer membrane proteins. *RNA*. 2008;14(3):543–551. https://doi.org/10.1261/rna.683308.
18. De Lay N, Gottesman S. Role of polynucleotide phosphorylase in sRNA function in Escherichia coli. *RNA*. 2011;17(6):1172–1189. https://doi.org/10.1261/rna.2531211.
19. Andrade JM, Pobre V, Matos AM, Arraiano CM. The crucial role of PNPase in the degradation of small RNAs that are not associated with Hfq. *RNA*. 2012;18(4): 844–855. https://doi.org/10.1261/rna.029413.111.
20. Li Z, Reimers S, Pandit S, Deutscher MP. RNA quality control: degradation of defective transfer RNA. *EMBO J*. 2002;21(5):1132–1138. https://doi.org/10.1093/emboj/21.5.1132.
21. Andrade JM, Cairrão F, Arraiano CM. RNase R affects gene expression in stationary phase: regulation of ompA. *Mol Microbiol*. 2006;60(1):219–228. https://doi.org/10.1111/j.1365-2958.2006.05092.x.
22. Régnier P, Hajnsdorf E. The interplay of Hfq, poly(A) polymerase I and exoribonucleases at the 3' ends of RNAs resulting from Rho-independent termination: a tentative model. *RNA Biol*. 2013;10(4):602–609. https://doi.org/10.4161/rna.23664.
23. Mohanty BK, Kushner SR. Polynucleotide phosphorylase functions both as a 3' right-arrow 5' exonuclease and a poly(A) polymerase in Escherichia coli. *Proc Natl Acad Sci USA*. 2000;97(22):11966–11971. https://doi.org/10.1073/pnas.220295997.
24. Grunberg-Manago M, Oritz PJ, Ochoa S. Enzymatic synthesis of nucleic acidlike polynucleotides. *Science*. 1955;122(3176):907–910.
25. Bermúdez-Cruz RM, Fernández-Ramírez F, Ramírez F, Kameyama-Kawabe L, Montañez C. Conserved domains in polynucleotide phosphorylase among eubacteria. *Biochimie*. 2005;87(8):737–745. https://doi.org/10.1016/j.biochi.2005.03.005.
26. Grunberg-Manago M. Enzymatic synthesis of nucleic acids. *Prog Biophys Mol Biol*. 1963;13:175–239.
27. Cardenas PP, Carrasco B, Sanchez H, Deikus G, Bechhofer DH, Alonso JC. Bacillus subtilis polynucleotide phosphorylase 3'-to-5' DNase activity is involved in DNA repair. *Nucleic Acids Res*. 2009;37(12):4157–4169. https://doi.org/10.1093/nar/gkp314.
28. Py B, Higgins CF, Krisch HM, Carpousis AJ. A DEAD-box RNA helicase in the Escherichia coli RNA degradosome. *Nature*. 1996;381(6578):169–172. https://doi.org/10.1038/381169a0.

29. Cheng Z-F, Deutscher MP. An important role for RNase R in mRNA decay. *Mol Cell.* 2005;17(2):313–318. https://doi.org/10.1016/j.molcel.2004.11.048.
30. Pobre V, Arraiano CM. Next generation sequencing analysis reveals that the ribonucleases RNase II, RNase R and PNPase affect bacterial motility and biofilm formation in E. coli. *BMC Genomics.* 2015;16(1):72. https://doi.org/10.1186/s12864-015-1237-6.
31. Andrade JM, Pobre V, Arraiano CM. Small RNA modules confer different stabilities and interact differently with multiple targets. In: Sumby P, ed. *PLoS One.* 2013;8(1):e52866. https://doi.org/10.1371/journal.pone.0052866.
32. Xu F, Cohen SN. RNA degradation in Escherichia coli regulated by 3' adenylation and 5' phosphorylation. *Nature.* 1995;374(6518):180–183. https://doi.org/10.1038/374180a0.
33. Spickler C, Mackie GA. Action of RNase II and polynucleotide phosphorylase against RNAs containing stem-loops of defined structure. *J Bacteriol.* 2000;182(9):2422–2427.
34. Godefroy T. Kinetics of polymerization and phosphorolysis reactions of Escherichia coli polynucleotide phosphorylase. Evidence for multiple binding of polynucleotide in phosphorolysis. *Eur J Biochem.* 1970;14(2):222–231.
35. Sulewski M, Marchese-Ragona SP, Johnson KA, Benkovic SJ. Mechanism of polynucleotide phosphorylase. *Biochemistry.* 1989;28(14):5855–5864.
36. Slomovic S, Portnoy V, Yehudai-Resheff S, Bronshtein E, Schuster G. Polynucleotide phosphorylase and the archaeal exosome as poly(A)-polymerases. *Biochim Biophys Acta.* 2008;1779(4):247–255. https://doi.org/10.1016/j.bbagrm.2007.12.004.
37. Mohanty BK, Kushner SR. The majority of Escherichia coli mRNAs undergo post-transcriptional modification in exponentially growing cells. *Nucleic Acids Res.* 2006;34(19):5695–5704. https://doi.org/10.1093/nar/gkl684.
38. Yehudai-Resheff S, Hirsh M, Schuster G. Polynucleotide phosphorylase functions as both an exonuclease and a poly(A) polymerase in spinach chloroplasts. *Mol Cell Biol.* 2001;21(16):5408–5416. https://doi.org/10.1128/MCB.21.16.5408-5416.2001.
39. Rott R, Zipor G, Portnoy V, Liveanu V, Schuster G. RNA polyadenylation and degradation in cyanobacteria are similar to the chloroplast but different from Escherichia coli. *J Biol Chem.* 2003;278(18):15771–15777. https://doi.org/10.1074/jbc.M211571200.
40. Sohlberg B, Huang J, Cohen SN. The Streptomyces coelicolor polynucleotide phosphorylase homologue, and not the putative poly(A) polymerase, can polyadenylate RNA. *J Bacteriol.* 2003;185(24):7273–7278.
41. Jones GH, Mackie GA. Streptomyces coelicolor polynucleotide phosphorylase can polymerize nucleoside diphosphates under phosphorolysis conditions, with implications for the degradation of structured RNAs. *J Bacteriol.* 2013;195(22):5151–5159. https://doi.org/10.1128/JB.00936-13.
42. Chou JY, Singer MF. Deoxyadenosine diphosphate as a substrate and inhibitor of polynucleotide phosphorylase of Micrococcus luteus. I. Deoxyadenosine diphosphate as a substrate for polymerization and the exchange reaction with inorganic ^{32}P. *J Biol Chem.* 1971;246(24):7486–7496.
43. Beljanski M. De novo synthesis of DNA-like molecules by polynucleotide phosphorylase in vitro. *J Mol Evol.* 1996;42(5):493–499.
44. Cardenas PP, Carzaniga T, Zangrossi S, et al. Polynucleotide phosphorylase exonuclease and polymerase activities on single-stranded DNA ends are modulated by RecN, SsbA and RecA proteins. *Nucleic Acids Res.* 2011;39(21):9250–9261. https://doi.org/10.1093/nar/gkr635.
45. Becket E, Tse L, Yung M, Cosico A, Miller JH. Polynucleotide phosphorylase plays an important role in the generation of spontaneous mutations in Escherichia coli. *J Bacteriol.* 2012;194(20):5613–5620. https://doi.org/10.1128/JB.00962-12.

46. Carzaniga T, Sbarufatti G, Briani F, Dehò G. Polynucleotide phosphorylase is implicated in homologous recombination and DNA repair in Escherichia coli. *BMC Microbiol.* 2017;17(1):81. https://doi.org/10.1186/s12866-017-0980-z.
47. Andrade JM, Dos Santos RF, Chelysheva I, Ignatova Z, Arraiano CM. The RNA-binding protein Hfq is important for ribosome biogenesis and affects translation fidelity. *EMBO J.* 2018;37(11):e97631. https://doi.org/10.15252/embj.201797631.
48. Mohanty BK, Maples VF, Kushner SR. The Sm-like protein Hfq regulates polyadenylation dependent mRNA decay in Escherichia coli. *Mol Microbiol.* 2004; 54(4):905–920. https://doi.org/10.1111/j.1365-2958.2004.04337.x.
49. Del Favero M, Mazzantini E, Briani F, Zangrossi S, Tortora P, Dehò G. Regulation of Escherichia coli polynucleotide phosphorylase by ATP. *J Biol Chem.* 2008;283(41): 27355–27359. https://doi.org/10.1074/jbc.C800113200.
50. Tuckerman JR, Gonzalez G, Gilles-Gonzalez M-A. Cyclic di-GMP activation of polynucleotide phosphorylase signal-dependent RNA processing. *J Mol Biol.* 2011;407(5): 633–639. https://doi.org/10.1016/j.jmb.2011.02.019.
51. Nurmohamed S, Vaidialingam B, Callaghan AJ, Luisi BF. Crystal structure of Escherichia coli polynucleotide phosphorylase core bound to RNase E, RNA and manganese: implications for catalytic mechanism and RNA degradosome assembly. *J Mol Biol.* 2009;389(1):17–33. https://doi.org/10.1016/j.jmb.2009.03.051.
52. Nurmohamed S, Vincent HA, Titman CM, et al. Polynucleotide phosphorylase activity may be modulated by metabolites in Escherichia coli. *J Biol Chem.* 2011;286(16): 14315–14323. https://doi.org/10.1074/jbc.M110.200741.
53. Wang G, Chen H-W, Oktay Y, et al. PNPASE regulates RNA import into mitochondria. *Cell.* 2010;142(3):456–467. https://doi.org/10.1016/j.cell.2010.06.035.
54. Stone CM, Butt LE, Bufton JC, et al. Inhibition of homologous phosphorolytic ribonucleases by citrate may represent an evolutionarily conserved communicative link between RNA degradation and central metabolism. *Nucleic Acids Res.* 2017;45(8): 4655–4666. https://doi.org/10.1093/nar/gkx114.
55. Portier C, Regnier P. Expression of the rpsO and pnp genes: structural analysis of a DNA fragment carrying their control regions. *Nucleic Acids Res.* 1984;12(15): 6091–6102.
56. Robert-Le Meur M, Portier C. E.coli polynucleotide phosphorylase expression is autoregulated through an RNase III-dependent mechanism. *EMBO J.* 1992;11(7): 2633–2641.
57. Robert-Le Meur M, Portier C. Polynucleotide phosphorylase of Escherichia coli induces the degradation of its RNase III processed messenger by preventing its translation. *Nucleic Acids Res.* 1994;22(3):397–403.
58. Jarrige AC, Mathy N, Portier C. PNPase autocontrols its expression by degrading a double-stranded structure in the pnp mRNA leader. *EMBO J.* 2001;20(23): 6845–6855. https://doi.org/10.1093/emboj/20.23.6845.
59. Carzaniga T, Dehò G, Briani F. RNase III-independent autogenous regulation of Escherichia coli polynucleotide phosphorylase via translational repression. In: Gourse RL, ed. *J Bacteriol.* 2015;197(11):1931–1938. https://doi.org/10.1128/JB.00105-15.
60. Mohanty BK, Kushner SR. Polyadenylation of Escherichia coli transcripts plays an integral role in regulating intracellular levels of polynucleotide phosphorylase and RNase E. *Mol Microbiol.* 2002;45(5):1315–1324.
61. Zilhão R, Cairrão F, Régnier P, Arraiano CM. PNPase modulates RNase II expression in Escherichia coli: implications for mRNA decay and cell metabolism. *Mol Microbiol.* 1996;20(5):1033–1042.
62. Park H, Yakhnin H, Connolly M, Romeo T, Babitzke P. CsrA participates in a PNPase autoregulatory mechanism by selectively repressing translation of pnp transcripts that

have been previously processed by RNase III and PNPase. In: Gourse RL, ed. *J Bacteriol.* 2015;197(24):3751–3759. https://doi.org/10.1128/JB.00721-15.
63. Vakulskas CA, Potts AH, Babitzke P, Ahmer BMM, Romeo T. Regulation of bacterial virulence by Csr (Rsm) systems. *Microbiol Mol Biol Rev.* 2015;79(2):193–224. https://doi.org/10.1128/MMBR.00052-14.
64. Piazza F, Zappone M, Sana M, Briani F, Dehò G. Polynucleotide phosphorylase of Escherichia coli is required for the establishment of bacteriophage P4 immunity. *J Bacteriol.* 1996;178(18):5513–5521.
65. Zangrossi S, Briani F, Ghisotti D, Regonesi ME, Tortora P, Dehò G. Transcriptional and post-transcriptional control of polynucleotide phosphorylase during cold acclimation in Escherichia coli. *Mol Microbiol.* 2000;36(6):1470–1480.
66. Jones PG, VanBogelen RA, Neidhardt FC. Induction of proteins in response to low temperature in Escherichia coli. *J Bacteriol.* 1987;169(5):2092–2095.
67. Mathy N, Jarrige AC, Robert-Le Meur M, Portier C. Increased expression of Escherichia coli polynucleotide phosphorylase at low temperatures is linked to a decrease in the efficiency of autocontrol. *J Bacteriol.* 2001;183(13):3848–3854. https://doi.org/10.1128/JB.183.13.3848-3854.2001.
68. Zhang Y, Burkhardt DH, Rouskin S, Li G-W, Weissman JS, Gross CA. A stress response that monitors and regulates mRNA structure is central to cold shock adaptation. *Mol Cell.* 2018;70(2):274–286.e7. https://doi.org/10.1016/j.molcel.2018.02.035.
69. Fontaine F, Gasiorowski E, Gracia C, et al. The small RNA SraG participates in PNPase homeostasis. *RNA.* 2016;22(10):1560–1573. https://doi.org/10.1261/rna.055236.115.
70. Miczak A, Kaberdin VR, Wei CL, Lin-Chao S. Proteins associated with RNase E in a multicomponent ribonucleolytic complex. *Proc Natl Acad Sci USA.* 1996;93(9):3865–3869.
71. Vanzo NF, Li YS, Py B, et al. Ribonuclease E organizes the protein interactions in the Escherichia coli RNA degradosome. *Genes Dev.* 1998;12(17):2770–2781.
72. Prud'homme-Généreux A, Beran RK, Iost I, Ramey CS, Mackie GA, Simons RW. Physical and functional interactions among RNase E, polynucleotide phosphorylase and the cold-shock protein, CsdA: evidence for a "cold shock degradosome". *Mol Microbiol.* 2004;54(5):1409–1421. https://doi.org/10.1111/j.1365-2958.2004.04360.x.
73. Hardwick SW, Chan VSY, Broadhurst RW, Luisi BF. An RNA degradosome assembly in Caulobacter crescentus. *Nucleic Acids Res.* 2011;39(4):1449–1459. https://doi.org/10.1093/nar/gkq928.
74. Lehnik-Habrink M, Pförtner H, Rempeters L, Pietack N, Herzberg C, Stülke J. The RNA degradosome in Bacillus subtilis: identification of CshA as the major RNA helicase in the multiprotein complex. *Mol Microbiol.* 2010;77(4):958–971. https://doi.org/10.1111/j.1365-2958.2010.07264.x.
75. Newman JA, Hewitt L, Rodrigues C, Solovyova AS, Harwood CR, Lewis RJ. Dissection of the network of interactions that links RNA processing with glycolysis in the Bacillus subtilis degradosome. *J Mol Biol.* 2012;416(1):121–136. https://doi.org/10.1016/j.jmb.2011.12.024.
76. Salvo E, Alabi S, Liu B, Schlessinger A, Bechhofer DH. Interaction of Bacillus subtilis polynucleotide phosphorylase and RNase Y: structural mapping and effect on mRNA turnover. *J Biol Chem.* 2016;291(13):6655–6663. https://doi.org/10.1074/jbc.M115.711044.
77. Cascante-Estepa N, Gunka K, Stülke J. Localization of components of the RNA-degrading machine in Bacillus subtilis. *Front Microbiol.* 2016;7:1492. https://doi.org/10.3389/fmicb.2016.01492.

78. Roux CM, DeMuth JP, Dunman PM. Characterization of components of the Staphylococcus aureus mRNA degradosome holoenzyme-like complex. *J Bacteriol.* 2011;193(19):5520–5526. https://doi.org/10.1128/JB.05485-11.
79. Wang X, Wang C, Wu M, et al. Enolase binds to RnpA in competition with PNPase in Staphylococcus aureus. *FEBS Lett.* 2017;591(21):3523–3535. https://doi.org/10.1002/1873-3468.12859.
80. Raj R, Mitra S, Gopal B. Characterization of Staphylococcus epidermidis polynucleotide phosphorylase and its interactions with ribonucleases RNase J1 and RNase J2. *Biochem Biophys Res Commun.* 2018;495(2):2078–2084. https://doi.org/10.1016/j.bbrc.2017.12.056.
81. Lin P-H, Lin-Chao S. RhlB helicase rather than enolase is the beta-subunit of the Escherichia coli polynucleotide phosphorylase (PNPase)-exoribonucleolytic complex. *Proc Natl Acad Sci USA.* 2005;102(46):16590–16595. https://doi.org/10.1073/pnas.0500994102.
82. Liou G-G, Chang H-Y, Lin C-S, Lin-Chao S. DEAD box RhlB RNA helicase physically associates with exoribonuclease PNPase to degrade double-stranded RNA independent of the degradosome-assembling region of RNase E. *J Biol Chem.* 2002;277(43):41157–41162. https://doi.org/10.1074/jbc.M206618200.
83. Tseng Y-T, Chiou N-T, Gogiraju R, Lin-Chao S. The Protein interaction of RNA helicase B (RhlB) and polynucleotide phosphorylase (PNPase) contributes to the homeostatic control of cysteine in Escherichia coli. *J Biol Chem.* 2015;290(50):29953–29963. https://doi.org/10.1074/jbc.M115.691881.
84. Portier C. Quaternary structure of polynucleotide phosphorylase from Escherichia coli: evidence of a complex between two types of polypeptide chains. *Eur J Biochem.* 1975;55(3):573–582.
85. Shi Z, Yang W-Z, Lin-Chao S, Chak K-F, Yuan HS. Crystal structure of Escherichia coli PNPase: central channel residues are involved in processive RNA degradation. *RNA.* 2008;14(11):2361–2371. https://doi.org/10.1261/rna.1244308.
86. Hardwick SW, Gubbey T, Hug I, Jenal U, Luisi BF. Crystal structure of Caulobacter crescentus polynucleotide phosphorylase reveals a mechanism of RNA substrate channelling and RNA degradosome assembly. *Open Biol.* 2012;2(4):120028. https://doi.org/10.1098/rsob.120028.
87. Symmons MF, Williams MG, Luisi BF, Jones GH, Carpousis AJ. Running rings around RNA: a superfamily of phosphate-dependent RNases. *Trends Biochem Sci.* 2002;27(1):11–18.
88. Mattaj IW. RNA recognition: a family matter? *Cell.* 1993;73(5):837–840.
89. Carzaniga T, Mazzantini E, Nardini M, et al. A conserved loop in polynucleotide phosphorylase (PNPase) essential for both RNA and ADP/phosphate binding. *Biochimie.* 2014;97:49–59. https://doi.org/10.1016/j.biochi.2013.09.018.
90. Jarrige A, Bréchemier-Baey D, Mathy N, Duché O, Portier C. Mutational analysis of polynucleotide phosphorylase from Escherichia coli. *J Mol Biol.* 2002;321(3):397–409.
91. Briani F, Del Favero M, Capizzuto R, et al. Genetic analysis of polynucleotide phosphorylase structure and functions. *Biochimie.* 2007;89(1):145–157. https://doi.org/10.1016/j.biochi.2006.09.020.
92. Yehudai-Resheff S, Portnoy V, Yogev S, Adir N, Schuster G. Domain analysis of the chloroplast polynucleotide phosphorylase reveals discrete functions in RNA degradation, polyadenylation, and sequence homology with exosome proteins. *Plant Cell.* 2003;15(9):2003–2019.
93. Sarkar D, Park ES, Emdad L, Randolph A, Valerie K, Fisher PB. Defining the domains of human polynucleotide phosphorylase (hPNPaseOLD-35) mediating cellular senescence. *Mol Cell Biol.* 2005;25(16):7333–7343. https://doi.org/10.1128/MCB.25.16.7333-7343.2005.

94. Matus-Ortega ME, Regonesi ME, Piña-Escobedo A, Tortora P, Dehò G, García-Mena J. The KH and S1 domains of Escherichia coli polynucleotide phosphorylase are necessary for autoregulation and growth at low temperature. *Biochim Biophys Acta.* 2007;1769(3):194–203. https://doi.org/10.1016/j.bbaexp.2007.01.008.
95. Wong AG, McBurney KL, Thompson KJ, Stickney LM, Mackie GA. S1 and KH domains of polynucleotide phosphorylase determine the efficiency of RNA binding and autoregulation. *J Bacteriol.* 2013;195(9):2021–2031. https://doi.org/10.1128/JB.00062-13.
96. Stickney LM, Hankins JS, Miao X, Mackie GA. Function of the conserved S1 and KH domains in polynucleotide phosphorylase. *J Bacteriol.* 2005;187(21):7214–7221. https://doi.org/10.1128/JB.187.21.7214-7221.2005.
97. Mohanty BK, Kushner SR. Genomic analysis in Escherichia coli demonstrates differential roles for polynucleotide phosphorylase and RNase II in mRNA abundance and decay. *Mol Microbiol.* 2003;50(2):645–658. https://doi.org/10.1046/j.1365-2958.2003.03724.x.
98. Bernstein JA, Lin P-H, Cohen SN, Lin-Chao S. Global analysis of Escherichia coli RNA degradosome function using DNA microarrays. *Proc Natl Acad Sci USA.* 2004;101(9):2758–2763. https://doi.org/10.1073/pnas.0308747101.
99. Liu B, Deikus G, Bree A, Durand S, Kearns DB, Bechhofer DH. Global analysis of mRNA decay intermediates in Bacillus subtilis wild-type and polynucleotide phosphorylase-deletion strains. *Mol Microbiol.* 2014;94(1):41–55. https://doi.org/10.1111/mmi.12748.
100. Liu B, Kearns DB, Bechhofer DH. Expression of multiple Bacillus subtilis genes is controlled by decay of slrA mRNA from Rho-dependent 3′ ends. *Nucleic Acids Res.* 2016;44(7):3364–3372. https://doi.org/10.1093/nar/gkw069.
101. Hayakawa H, Kuwano M, Sekiguchi M. Specific binding of 8-oxoguanine-containing RNA to polynucleotide phosphorylase protein. *Biochemistry.* 2001;40(33):9977–9982.
102. Khidr L, Wu G, Davila A, Procaccio V, Wallace D, Lee WH. Role of SUV3 helicase in maintaining mitochondrial homeostasis in human cells. *J Biol Chem.* 2008;283(40):27064–27073. https://doi.org/10.1074/jbc.M802991200.
103. Wu J, Jiang Z, Liu M, et al. Polynucleotide phosphorylase protects Escherichia coli against oxidative stress. *Biochemistry.* 2009;48(9):2012–2020. https://doi.org/10.1021/bi801752p.
104. Reuven NB, Zhou Z, Deutscher MP. Functional overlap of tRNA nucleotidyltransferase, poly(A) polymerase I, and polynucleotide phosphorylase. *J Biol Chem.* 1997;272(52):33255–33259. https://doi.org/10.1074/jbc.272.52.33255.
105. Maes A, Gracia C, Hajnsdorf E, Régnier P. Search for poly(A) polymerase targets in E. coli reveals its implication in surveillance of Glu tRNA processing and degradation of stable RNAs. *Mol Microbiol.* 2012;83(2):436–451. https://doi.org/10.1111/j.1365-2958.2011.07943.x.
106. Mohanty BK, Petree JR, Kushner SR. Endonucleolytic cleavages by RNase E generate the mature 3′ termini of the three proline tRNAs in Escherichia coli. *Nucleic Acids Res.* 2016;44(13):6350–6362. https://doi.org/10.1093/nar/gkw517.
107. Mohanty BK, Kushner SR. Processing of the Escherichia coli leuX tRNA transcript, encoding tRNA(Leu5), requires either the 3'→5' exoribonuclease polynucleotide phosphorylase or RNase P to remove the Rho-independent transcription terminator. *Nucleic Acids Res.* 2010;38(2):597–607. https://doi.org/10.1093/nar/gkp997.
108. Sulthana S, Deutscher MP. Multiple exoribonucleases catalyze maturation of the 3' terminus of 16S ribosomal RNA (rRNA). *J Biol Chem.* 2013;288(18):12574–12579. https://doi.org/10.1074/jbc.C113.459172.
109. Smith BA, Gupta N, Denny K, Culver GM. Characterization of 16S rRNA processing with Pre-30S subunit assembly intermediates from E. coli. *J Mol Biol.* 2018;430(12):1745–1759. https://doi.org/10.1016/j.jmb.2018.04.009.

110. Wurtmann EJ, Wolin SL. A role for a bacterial ortholog of the Ro autoantigen in starvation-induced rRNA degradation. *Proc Natl Acad Sci USA*. 2010;107(9): 4022–4027. https://doi.org/10.1073/pnas.1000307107.
111. Chen X, Taylor DW, Fowler CC, Galan JE, Wang H-W, Wolin SL. An RNA degradation machine sculpted by Ro autoantigen and noncoding RNA. *Cell*. 2013;153(1): 166–177. https://doi.org/10.1016/j.cell.2013.02.037.
112. Chen X, Sim S, Wurtmann EJ, Feke A, Wolin SL. Bacterial noncoding Y RNAs are widespread and mimic tRNAs. *RNA*. 2014;20(11):1715–1724. https://doi.org/10.1261/rna.047241.114.
113. Viegas SC, Pfeiffer V, Sittka A, Silva IJ, Vogel J, Arraiano CM. Characterization of the role of ribonucleases in Salmonella small RNA decay. *Nucleic Acids Res*. 2007;35(22):7651–7664. https://doi.org/10.1093/nar/gkm916.
114. Vogel J, Luisi BF. Hfq and its constellation of RNA. *Nat Rev Microbiol*. 2011;9(8):578–589. https://doi.org/10.1038/nrmicro2615.
115. Das SK, Sokhi UK, Bhutia SK, et al. Human polynucleotide phosphorylase selectively and preferentially degrades microRNA-221 in human melanoma cells. *Proc Natl Acad Sci USA*. 2010;107(26):11948–11953. https://doi.org/10.1073/pnas.0914143107.
116. Sesto N, Touchon M, Andrade JM, et al. A PNPase dependent CRISPR system in listeria. In: Burkholder WF, ed. *PLoS Genet*. 2014;10(1):e1004065. https://doi.org/10.1371/journal.pgen.1004065.
117. Cameron TA, De Lay NR. The phosphorolytic exoribonucleases polynucleotide phosphorylase and RNase PH stabilize sRNAs and facilitate regulation of their mRNA targets. *J Bacteriol*. 2016;198(24):3309–3317. https://doi.org/10.1128/JB.00624-16.
118. Bandyra KJ, Sinha D, Syrjanen J, Luisi BF, De Lay NR. The ribonuclease polynucleotide phosphorylase can interact with small regulatory RNAs in both protective and degradative modes. *RNA*. 2016;22(3):360–372. https://doi.org/10.1261/rna.052886.115.
119. Schuster G, Stern D. RNA polyadenylation and decay in mitochondria and chloroplasts. *Prog Mol Biol Transl Sci*. 2009;85:393–422. https://doi.org/10.1016/S0079-6603(08)00810-6.
120. Wang DD-H, Shu Z, Lieser SA, Chen P-L, Lee W-H. Human mitochondrial SUV3 and polynucleotide phosphorylase form a 330-kDa heteropentamer to cooperatively degrade double-stranded RNA with a 3'-to-5' directionality. *J Biol Chem*. 2009;284(31):20812–20821. https://doi.org/10.1074/jbc.M109.009605.
121. Germain A, Kim SH, Gutierrez R, Stern DB. Ribonuclease II preserves chloroplast RNA homeostasis by increasing mRNA decay rates, and cooperates with polynucleotide phosphorylase in 3′ end maturation. *Plant J*. 2012;72(6):960–971. https://doi.org/10.1111/tpj.12006.
122. Castandet B, Hotto AM, Fei Z, Stern DB. Strand-specific RNA sequencing uncovers chloroplast ribonuclease functions. *FEBS Lett*. 2013;587(18):3096–3101. https://doi.org/10.1016/j.febslet.2013.08.004.
123. Wang DD-H, Guo XE, Modrek AS, Chen C-F, Chen P-L, Lee W-H. Helicase SUV3, polynucleotide phosphorylase, and mitochondrial polyadenylation polymerase form a transient complex to modulate mitochondrial mRNA polyadenylated tail lengths in response to energetic changes. *J Biol Chem*. 2014;289(24):16727–16735. https://doi.org/10.1074/jbc.M113.536540.
124. Toompuu M, Tuomela T, Laine P, Paulin L, Dufour E, Jacobs HT. Polyadenylation and degradation of structurally abnormal mitochondrial tRNAs in human cells. *Nucleic Acids Res*. 2018;46(10):5209–5226. https://doi.org/10.1093/nar/gky159.
125. Chen H-W, Rainey RN, Balatoni CE, et al. Mammalian polynucleotide phosphorylase is an intermembrane space RNase that maintains mitochondrial homeostasis. *Mol Cell Biol*. 2006;26(22):8475–8487. https://doi.org/10.1128/MCB.01002-06.

126. Liu X, Fu R, Pan Y, Meza-Sosa KF, Zhang Z, Lieberman J. PNPT1 release from mitochondria during apoptosis triggers decay of poly(A) RNAs. *Cell*. 2018;174(1):187–201.e12. https://doi.org/10.1016/j.cell.2018.04.017.
127. Leszczyniecka M, Kang DC, Sarkar D, et al. Identification and cloning of human polynucleotide phosphorylase, hPNPaseold-35, in the context of terminal differentiation and cellular senescence. *Proc Natl Acad Sci USA*. 2002;99(26):16636–16641. https://doi.org/10.1073/pnas.252643699.
128. Leszczyniecka M, Su Z, Kang D, Sarkar D, Fisher PB. Expression regulation and genomic organization of human polynucleotide phosphorylase, hPNPaseold-35, a type I interferon inducible early response gene. *Gene*. 2003;316:143–156. https://doi.org/10.1016/S0378-1119(03)00752-2.
129. Sarkar D, Leszczyniecka M, Kang D, et al. Down-regulation of Myc as a potential target for growth arrest induced by human polynucleotide phosphorylase (hPNPaseold-35) in human melanoma cells. *J Biol Chem*. 2003;278(27):24542–24551. https://doi.org/10.1074/jbc.M302421200.
130. Sarkar D, Park ES, Fisher PB. Defining the mechanism by which IFN-beta dowregulates c-myc expression in human melanoma cells: pivotal role for human polynucleotide phosphorylase (hPNPaseold-35). *Cell Death Differ*. 2006;13(9):1541–1553. https://doi.org/10.1038/sj.cdd.4401829.
131. Sarkar D, Fisher PB. Molecular mechanisms of aging-associated inflammation. *Cancer Lett*. 2006;236(1):13–23. https://doi.org/10.1016/j.canlet.2005.04.009.
132. Wu J, Li Z. Human polynucleotide phosphorylase reduces oxidative RNA damage and protects HeLa cell against oxidative stress. *Biochem Biophys Res Commun*. 2008;372(2):288–292. https://doi.org/10.1016/j.bbrc.2008.05.058.
133. Lin CL, Wang Y-T, Yang W-Z, Hsiao Y-Y, Yuan HS. Crystal structure of human polynucleotide phosphorylase: insights into its domain function in RNA binding and degradation. *Nucleic Acids Res*. 2012;40(9):4146–4157. https://doi.org/10.1093/nar/gkr1281.
134. Barnett TC, Bugrysheva JV, Scott JR. Role of mRNA stability in growth phase regulation of gene expression in the group A streptococcus. *J Bacteriol*. 2007;189(5):1866–1873. https://doi.org/10.1128/JB.01658-06.
135. Rosenzweig JA, Weltman G, Plano GV, Schesser K. Modulation of yersinia type three secretion system by the S1 domain of polynucleotide phosphorylase. *J Biol Chem*. 2005;280(1):156–163. https://doi.org/10.1074/jbc.M405662200.
136. Rosenzweig JA, Chromy B, Echeverry A, et al. Polynucleotide phosphorylase independently controls virulence factor expression levels and export in Yersinia spp. *FEMS Microbiol Lett*. 2007;270(2):255–264. https://doi.org/10.1111/j.1574-6968.2007.00689.x.
137. Chen R, Weng Y, Zhu F, et al. Polynucleotide phosphorylase regulates multiple virulence factors and the stabilities of small RNAs RsmY/Z in Pseudomonas aeruginosa. *Front Microbiol*. 2016;7. https://doi.org/10.3389/fmicb.2016.00247.
138. Hu J, Zhu MJ. Defects in polynucleotide phosphorylase impairs virulence in Escherichia coli O157:H7. *Front Microbiol*. 2015;6(JUL):1–9. https://doi.org/10.3389/fmicb.2015.00806.
139. Engman J, Negrea A, Sigurlásdóttir S, et al. Neisseria meningitidis polynucleotide phosphorylase affects aggregation, adhesion, and virulence. In: Payne SM, ed. *Infect Immun*. 2016;84(5):1501–1513. https://doi.org/10.1128/IAI.01463-15.
140. Haddad N, Tresse O, Rivoal K, et al. Polynucleotide phosphorylase has an impact on cell biology of Campylobacter jejuni. *Front Cell Infect Microbiol*. 2012;2:30. https://doi.org/10.3389/fcimb.2012.00030.
141. Clements MO, Eriksson S, Thompson A, et al. Polynucleotide phosphorylase is a global regulator of virulence and persistency in Salmonella enterica. *Proc Natl Acad Sci USA*. 2002;99(13):8784–8789. https://doi.org/10.1073/pnas.132047099.

142. Bearson SMD, Bearson BL, Lee IS, Kich JD. Polynucleotide phosphorylase (PNPase) is required for Salmonella enterica serovar Typhimurium colonization in swine. *Microb Pathog.* 2013;65:63–66. https://doi.org/10.1016/j.micpath.2013.10.001.
143. Carzaniga T, Antoniani D, Dehò G, Briani F, Landini P. The RNA processing enzyme polynucleotide phosphorylase negatively controls biofilm formation by repressing poly-N-acetylglucosamine (PNAG) production in Escherichia coli C. *BMC Microbiol.* 2012;12(1):270. https://doi.org/10.1186/1471-2180-12-270.
144. Rainey RN, Glavin JD, Chen H-W, French SW, Teitell MA, Koehler CM. A new function in translocation for the mitochondrial i-AAA protease Yme1: import of polynucleotide phosphorylase into the intermembrane space. *Mol Cell Biol.* 2006;26(22):8488–8497. https://doi.org/10.1128/MCB.01006-06.
145. Wang G, Shimada E, Koehler CM, Teitell MA. PNPASE and RNA trafficking into mitochondria. *Biochim Biophys Acta.* 2012;1819(9-10):998–1007. https://doi.org/10.1016/j.bbagrm.2011.10.001.
146. Wang G, Shimada E, Zhang J, et al. Correcting human mitochondrial mutations with targeted RNA import. *Proc Natl Acad Sci USA.* 2012;109(13):4840–4845. https://doi.org/10.1073/pnas.1116792109.
147. Matilainen S, Carroll CJ, Richter U, et al. Defective mitochondrial RNA processing due to PNPT1 variants causes Leigh syndrome. *Hum Mol Genet.* 2017;26(17):3352–3361. https://doi.org/10.1093/hmg/ddx221.
148. Sato R, Arai-Ichinoi N, Kikuchi A, et al. Novel biallelic mutations in the PNPT1 gene encoding a mitochondrial-RNA-import protein PNPase cause delayed myelination. *Clin Genet.* 2018;93(2):242–247. https://doi.org/10.1111/cge.13068.
149. Shepherd DL, Hathaway QA, Pinti MV, et al. Exploring the mitochondrial microRNA import pathway through polynucleotide phosphorylase (PNPase). *J Mol Cell Cardiol.* 2017;110:15–25. https://doi.org/10.1016/j.yjmcc.2017.06.012.
150. Marujo PE, Hajnsdorf E, Le Derout J, Andrade R, Arraiano CM, Régnier P. RNase II removes the oligo(A) tails that destabilize the rpsO mRNA of Escherichia coli. *RNA.* 2000;6(8):1185–1193.
151. Cannistraro VJ, Kennell D. The reaction mechanism of ribonuclease II and its interaction with nucleic acid secondary structures. *Biochim Biophys Acta Protein Struct Mol Enzymol.* 1999;1433(1-2):170–187. https://doi.org/10.1016/S0167-4838(99)00136-3.
152. Deutscher MP, Reuven NB. Enzymatic basis for hydrolytic versus phosphorolytic mRNA degradation in Escherichia coli and Bacillus subtilis. *Proc Natl Acad Sci USA.* 1991;88(8):3277–3280.
153. Mohanty BK, Kushner SR. Polynucleotide phosphorylase, RNase II ansi RNase E play different roles in the in vivo modulation of polyadenylation in Escherichia coli. *Mol Microbiol.* 2000;36(4):982–994. https://doi.org/10.1046/j.1365-2958.2000.01921.x.
154. Dziembowski A, Lorentzen E, Conti E, Séraphin B. A single subunit, Dis3, is essentially responsible for yeast exosome core activity. *Nat Struct Mol Biol.* 2007;14(1):15–22. https://doi.org/10.1038/nsmb1184.
155. Tomecki R, Kristiansen MS, Lykke-Andersen S, et al. The human core exosome interacts with differentially localized processive RNases: HDIS3 and hDIS3L. *EMBO J.* 2010;29(14):2342–2357. https://doi.org/10.1038/emboj.2010.121.
156. Staals RHJ, Bronkhorst AW, Schilders G, et al. Dis3-like 1: a novel exoribonuclease associated with the human exosome. *EMBO J.* 2010;29(14):2358–2367. https://doi.org/10.1038/emboj.2010.122.
157. Łabno A, Warkocki Z, Kuliński T, et al. Perlman syndrome nuclease DIS3L2 controls cytoplasmic non-coding RNAs and provides surveillance pathway for maturing snRNAs. *Nucleic Acids Res.* 2016;44(21):10437–10453. https://doi.org/10.1093/nar/gkw649.

158. Song L, Wang G, Malhotra A, Deutscher MP, Liang W. Reversible acetylation on Lys501 regulates the activity of RNase II. *Nucleic Acids Res*. 2016;44(5):1979–1988. https://doi.org/10.1093/nar/gkw053.
159. Liang W, Malhotra A, Deutscher MP. Acetylation regulates the stability of a bacterial protein: growth stage-dependent modification of RNase R. *Mol Cell*. 2011;44(1):160–166. https://doi.org/10.1016/j.molcel.2011.06.037.
160. Zilhão R, Plumbridge J, Hajnsdorf E, Régnier P, Arraiano CM. Escherichia coli RNase II: characterization of the promoters involved in the transcription of rnb. *Microbiology*. 1996;142(2):367–375. https://doi.org/10.1099/13500872-142-2-367.
161. Sulthana S, Quesada E, Deutscher MP. RNase II regulates RNase PH and is essential for cell survival during starvation and stationary phase. *RNA*. 2017;23(9):1456–1464. https://doi.org/10.1261/rna.060558.116.
162. Zilhão R, Régnier P, Maria AC. The role of endonucleases in the expression of ribonuclease II in Escherichia coli. *FEMS Microbiol Lett*. 1995;130(2–3):237–244. https://doi.org/10.1016/0378-1097(95)00212-N.
163. Cairrão F, Chora Â, Zilhão R, Carpousis AJ, Arraiano CM. RNase II levels change according to the growth conditions: characterization of gmr, a new Escherichia coli gene involved in the modulation of RNase II. *Mol Microbiol*. 2001;39(6):1550–1561. https://doi.org/10.1046/j.1365-2958.2001.02342.x.
164. Mitchell P, Petfalski E, Shevchenko A, Mann M, Tollervey D. The exosome: a conserved eukaryotic RNA processing complex containing multiple 3'→5' exoribonucleases. *Cell*. 1997;91(4):457–466. https://doi.org/10.1016/S0092-8674(00)80432-8.
165. Liu Q, Greimann JC, Lima CD. Reconstitution, activities, and structure of the Eukaryotic RNA exosome. *Cell*. 2006;127(6):1223–1237. https://doi.org/10.1016/j.cell.2006.10.037.
166. Lorentzen E, Basquin J, Tomecki R, Dziembowski A, Conti E. Structure of the active subunit of the yeast exosome core, Rrp44: diverse modes of substrate recruitment in the RNase II nuclease family. *Mol Cell*. 2008;29(6):717–728. https://doi.org/10.1016/j.molcel.2008.02.018.
167. Ibrahim A, Marbán E. Exosomes: fundamental biology and roles in cardiovascular physiology. *Annu Rev Physiol*. 2016;78(1):67–83. https://doi.org/10.1146/annurev-physiol-021115-104929.
168. Lu F, Taghbalout A. The Escherichia coli major exoribonuclease RNase II is a component of the RNA degradosome. *Biosci Rep*. 2014;34(6):e00166. https://doi.org/10.1042/BSR20140113.
169. Zuo Y, Vincent HA, Zhang J, Wang Y, Deutscher MP, Malhotra A. Structural basis for processivity and single-strand specificity of RNase II. *Mol Cell*. 2006;24(1):149–156. https://doi.org/10.1016/j.molcel.2006.09.004.
170. Lebreton A, Tomecki R, Dziembowski A, Séraphin B. Endonucleolytic RNA cleavage by a eukaryotic exosome. *Nature*. 2008;456(7224):993–996. https://doi.org/10.1038/nature07480.
171. Schaeffer D, Tsanova B, Barbas A, et al. The exosome contains domains with specific endoribonuclease, exoribonuclease and cytoplasmic mRNA decay activities. *Nat Struct Mol Biol*. 2009;16(1):56–62. https://doi.org/10.1038/nsmb.1528.
172. Schneider C, Leung E, Brown J, Tollervey D. The N-terminal PIN domain of the exosome subunit Rrp44 harbors endonuclease activity and tethers Rrp44 to the yeast core exosome. *Nucleic Acids Res*. 2009;37(4):1127–1140. https://doi.org/10.1093/nar/gkn1020.
173. Barbas A, Matos RG, Amblar M, López-Viñas E, Gomez-Puertas P, Arraiano CM. New insights into the mechanism of RNA degradation by ribonuclease II: identification of the residue responsible for setting the RNase II end product. *J Biol Chem*. 2008;283(19):13070–13076. https://doi.org/10.1074/jbc.M709989200.

174. Barbas A, Matos RG, Amblar M, López-Viñas E, Gomez-Puertas P, Arraiano CM. Determination of key residues for catalysis and RNA cleavage specificity. One mutation turns RNase II Into a "super-enzyme". *J Biol Chem*. 2009;284(31):20486–20498. https://doi.org/10.1074/jbc.M109.020693.
175. Lu F, Taghbalout A. Membrane association via an amino-terminal amphipathic helix is required for the cellular organization and function of RNase II. *J Biol Chem*. 2013;288(10):7241–7251. https://doi.org/10.1074/jbc.M112.408674.
176. Schmier BJ, Seetharaman J, Deutscher MP, Hunt JF, Malhotra A. The structure and enzymatic properties of a novel RNase II family enzyme from Deinococcus radiodurans. *J Mol Biol*. 2012;415(3):547–559. https://doi.org/10.1016/j.jmb.2011.11.031.
177. Janssen B, Hayes C. The tmRNA ribosome rescue system. *Adv Protein Chem Struct Biol*. 2012;86:151–159. https://doi.org/10.1016/B978-0-12-386497-0.00005-0.
178. Janssen BD, Garza-Sánchez F, Hayes CS. A-site mRNA cleavage is not required for tmRNA-mediated ssrA-peptide tagging. *PLoS One*. 2013;8(11). https://doi.org/10.1371/journal.pone.0081319.
179. Li Z, Deutscher MP. Maturation pathways for E. coli tRNA precursors: a random multienzyme process in vivo. *Cell*. 1996;86(3):503–512.
180. Li Z, Pandit S, Deutscher MP. 3' exoribonucleolytic trimming is a common feature of the maturation of small, stable RNAs in Escherichia coli. *Proc Natl Acad Sci USA*. 1998;95(6):2856–2861. https://doi.org/10.1073/pnas.95.6.2856.
181. Basturea GN, Zundel MA, Deutscher MP. Degradation of ribosomal RNA during starvation: comparison to quality control during steady-state growth and a role for RNase PH. *RNA*. 2011;17(2):338–345. https://doi.org/10.1261/rna.2448911.
182. Jacob AI, Köhrer C, Davies BW, RajBhandary UL, Walker GC. Conserved bacterial RNase YbeY plays key roles in 70S ribosome quality control and 16S rRNA maturation. *Mol Cell*. 2013;49(3):427–438. https://doi.org/10.1016/j.molcel.2012.11.025.
183. Cheng Z-FF, Deutscher MP. Purification and characterization of the Escherichia coli exoribonuclease RNase R. Comparison with RNase II. *J Biol Chem*. 2002;277(24):21624–21629. https://doi.org/10.1074/jbc.M202942200.
184. Vincent HA, Deutscher MP. Substrate recognition and catalysis by the exoribonuclease RNase R. *J Biol Chem*. 2006;281(40):29769–29775. https://doi.org/10.1074/jbc.M606744200.
185. Hossain ST, Malhotra A, Deutscher MP. How RNase R degrades structured RNA: role of the helicase activity and the S1 domain. *J Biol Chem*. 2016;291(15):7877–7887. https://doi.org/10.1074/jbc.M116.717991.
186. Ebbesen KK, Kjems J, Hansen TB. Circular RNAs: identification, biogenesis and function. *Biochim Biophys Acta*. 2016;1859(1):163–168. https://doi.org/10.1016/j.bbagrm.2015.07.007.
187. Suzuki H, Zuo Y, Wang J, Zhang MQ, Malhotra A, Mayeda A. Characterization of RNase R-digested cellular RNA source that consists of lariat and circular RNAs from pre-mRNA splicing. *Nucleic Acids Res*. 2006;34(8). , e63https://doi.org/10.1093/nar/gkl151.
188. Suzuki H, Tsukahara T. A view of pre-mRNA splicing from RNase R resistant RNAs. *Int J Mol Sci*. 2014;15(6):9331–9342. https://doi.org/10.3390/ijms15069331.
189. Cairrão F, Arraiano CM. The role of endoribonucleases in the regulation of RNase R. *Biochem Biophys Res Commun*. 2006;343(3):731–737. https://doi.org/10.1016/j.bbrc.2006.03.040.
190. Cairrão F, Cruz A, Mori H, Arraiano CM. Cold shock induction of RNase R and its role in the maturation of the quality control mediator SsrA/tmRNA. *Mol Microbiol*. 2003;50(4):1349–1360.
191. Chen C, Deutscher MP. Elevation of RNase R in response to multiple stress conditions. *J Biol Chem*. 2005;280(41):34393–34396. https://doi.org/10.1074/jbc.C500333200.

192. Chen C, Deutscher MP. RNase R is a highly unstable protein regulated by growth phase and stress. *RNA*. 2010;16(4):667–672. https://doi.org/10.1261/rna.1981010.
193. Liang W, Deutscher MP. Post-translational modification of RNase R is regulated by stress-dependent reduction in the acetylating enzyme Pka (YfiQ). *RNA*. 2012;18(1):37–41. https://doi.org/10.1261/rna.030213.111.
194. Liang W, Deutscher MP. Transfer-messenger RNA-SmpB protein regulates ribonuclease R turnover by promoting binding of HslUV and Lon proteases. *J Biol Chem*. 2012;287(40):33472–33479. https://doi.org/10.1074/jbc.M112.375287.
195. Liang W, Deutscher MP. Ribosomes regulate the stability and action of the exoribonuclease RNase R. *J Biol Chem*. 2013;288(48):34791–34798. https://doi.org/10.1074/jbc.M113.519553.
196. Purusharth RI, Madhuri B, Ray MK. Exoribonuclease R in Pseudomonas syringae is essential for growth at low temperature and plays a novel role in the 3' end processing of 16 and 5 S ribosomal RNA. *J Biol Chem*. 2007;282(22):16267–16277. https://doi.org/10.1074/jbc.M605588200.
197. Carabetta VJ, Silhavy TJ, Cristea IM. The response regulator SprE (RssB) is required for maintaining poly(A) polymerase I-degradosome association during stationary phase. *J Bacteriol*. 2010;192(14):3713–3721. https://doi.org/10.1128/JB.00300-10.
198. Strader MB, Hervey WJ, Costantino N, et al. A coordinated proteomic approach for identifying proteins that interact with the E. coli ribosomal protein S12. *J Proteome Res*. 2013;12(3):1289–1299. https://doi.org/10.1021/pr3009435.
199. Hossain ST, Malhotra A, Deutscher MP. The helicase activity of ribonuclease R is essential for efficient nuclease activity. *J Biol Chem*. 2015;290(25):15697–15706. https://doi.org/10.1074/jbc.M115.650176.
200. Chu L-Y, Hsieh T-J, Golzarroshan B, Chen Y-P, Agrawal S, Yuan HS. Structural insights into RNA unwinding and degradation by RNase R. *Nucleic Acids Res*. 2017;45(20):12015–12024. https://doi.org/10.1093/nar/gkx880.
201. Matos RG, Barbas A, Arraiano CM. RNase R mutants elucidate the catalysis of structured RNA: RNA-binding domains select the RNAs targeted for degradation. *Biochem J*. 2009;423(2):291–301. https://doi.org/10.1042/BJ20090839.
202. Vincent HA, Deutscher MP. Insights into how RNase R degrades structured RNA: analysis of the nuclease domain. *J Mol Biol*. 2009;387(3):570–583. https://doi.org/10.1016/j.jmb.2009.01.068.
203. Faehnle CR, Walleshauser J, Joshua-Tor L. Mechanism of Dis3l2 substrate recognition in the Lin28-let-7 pathway. *Nature*. 2014;514(7521):252–256. https://doi.org/10.1038/nature13553.
204. Vuković L, Chipot C, Makino DL, Conti E, Schulten K. Molecular mechanism of processive 3′ to 5′ RNA translocation in the active subunit of the RNA exosome complex. *J Am Chem Soc*. 2016. https://doi.org/10.1021/jacs.5b12065.
205. Hammarlöf DL, Bergman JM, Garmendia E, Hughes D. Turnover of mRNAs is one of the essential functions of RNase E. *Mol Microbiol*. 2015;98(1):34–45. https://doi.org/10.1111/mmi.13100.
206. Andrade JM, Hajnsdorf E, Régnier P, Arraiano CM. The poly(A)-dependent degradation pathway of rpsO mRNA is primarily mediated by RNase R. *RNA*. 2009;15(2):316–326. https://doi.org/10.1261/rna.1197309.
207. Phadtare S, Severinov K. RNA remodeling and gene regulation by cold shock proteins. *RNA Biol*. 2010;7(6):788–795.
208. Söderberg MA, Cianciotto NP. Mediators of lipid A modification, RNA degradation, and central intermediary metabolism facilitate the growth of Legionella pneumophila at low temperatures. *Curr Microbiol*. 2010;60(1):59–65. https://doi.org/10.1007/s00284-009-9502-z.

209. Charpentier X, Faucher SP, Kalachikov S, Shuman HA. Loss of RNase R induces competence development in Legionella pneumophila. *J Bacteriol.* 2008;190(24): 8126–8136. https://doi.org/10.1128/JB.01035-08.
210. Awano N, Xu C, Ke H, Inoue K, Inouye M, Phadtare S. Complementation analysis of the cold-sensitive phenotype of the Escherichia coli csdA deletion strain. *J Bacteriol.* 2007;189(16):5808–5815. https://doi.org/10.1128/JB.00655-07.
211. Phadtare S. Escherichia coli cold-shock gene profiles in response to over-expression/deletion of CsdA, RNase R and PNPase and relevance to low-temperature RNA metabolism. *Genes Cells.* 2012;17(10):850–874. https://doi.org/10.1111/gtc.12002.
212. Zundel MA, Basturea GN, Deutscher MP. Initiation of ribosome degradation during starvation in Escherichia coli. *RNA.* 2009;15(5):977–983. https://doi.org/10.1261/rna.1381309.
213. Kaplan R, Apirion D. The fate of ribosomes in Escherichia coli cells starved for a carbon source. *J Biol Chem.* 1975;250(5):1854–1863.
214. Davies BW, Köhrer C, Jacob AI, et al. Role of Escherichia coli YbeY, a highly conserved protein, in rRNA processing. *Mol Microbiol.* 2010;78(2):506–518. https://doi.org/10.1111/j.1365-2958.2010.07351.x.
215. Bollenbach TJ, Lange H, Gutierrez R, Erhardt M, Stern DB, Gagliardi D. RNR1, a 3'-5' exoribonuclease belonging to the RNR superfamily, catalyzes 3' maturation of chloroplast ribosomal RNAs in Arabidopsis thaliana. *Nucleic Acids Res.* 2005;33(8):2751–2763. https://doi.org/10.1093/nar/gki576.
216. Wellner K, Betat H, Mörl M. A tRNA's fate is decided at its 3' end: collaborative actions of CCA-adding enzyme and RNases involved in tRNA processing and degradation. *Biochim Biophys Acta.* 2018;1861(4):433–441. https://doi.org/10.1016/j.bbagrm.2018.01.012.
217. Campos-Guillén J, Arvizu-Gómez JL, Jones GH, Olmedo-Alvarez G. Characterization of tRNA(Cys) processing in a conditional Bacillus subtilis CCase mutant reveals the participation of RNase R in its quality control. *Microbiology.* 2010;156(Pt 7): 2102–2111. https://doi.org/10.1099/mic.0.034652-0.
218. Alluri RK, Li Z. Novel one-step mechanism for tRNA 3'-end maturation by the exoribonuclease RNase R of Mycoplasma genitalium. *J Biol Chem.* 2012;287(28):23427–23433. https://doi.org/10.1074/jbc.M111.324970.
219. Russell JH, Keiler KC. Subcellular localization of a bacterial regulatory RNA. *Proc Natl Acad Sci USA.* 2009;106(38):16405–16409. https://doi.org/10.1073/pnas.0904904106.
220. Karzai AW, Sauer RT. Protein factors associated with the SsrA.SmpB tagging and ribosome rescue complex. *Proc Natl Acad Sci USA.* 2001;98(6):3040–3044. https://doi.org/10.1073/pnas.051628298.
221. Venkataraman K, Guja KE, Garcia-Diaz M, Karzai AW. Non-stop mRNA decay: a special attribute of trans-translation mediated ribosome rescue. *Front Microbiol.* 2014;5:93. https://doi.org/10.3389/fmicb.2014.00093.
222. Richards J, Mehta P, Karzai AW. RNase R degrades non-stop mRNAs selectively in an SmpB-tmRNA-dependent manner. *Mol Microbiol.* 2006;62(6):1700–1712. https://doi.org/10.1111/j.1365-2958.2006.05472.x.
223. Ge Z, Mehta P, Richards J, Karzai AW. Non-stop mRNA decay initiates at the ribosome. *Mol Microbiol.* 2010;78(5):1159–1170. https://doi.org/10.1111/j.1365-2958.2010.07396.x.
224. Jahn N, Preis H, Wiedemann C, Brantl S. BsrG/SR4 from Bacillus subtilis—the first temperature-dependent type I toxin-antitoxin system. *Mol Microbiol.* 2012;83(3): 579–598. https://doi.org/10.1111/j.1365-2958.2011.07952.x.
225. Zhang W, Murphy C, Sieburth LE. Conserved RNaseII domain protein functions in cytoplasmic mRNA decay and suppresses Arabidopsis decapping mutant phenotypes.

Proc Natl Acad Sci USA. 2010;107(36):15981–15985. https://doi.org/10.1073/pnas.1007060107.
226. Arraiano CM, Andrade JM, Domingues S, et al. The critical role of RNA processing and degradation in the control of gene expression. *FEMS Microbiol Rev*. 2010;34(5):883–923. https://doi.org/10.1111/j.1574-6976.2010.00242.x.
227. Schneider C, Anderson JT, Tollervey D. The exosome subunit Rrp44 plays a direct role in RNA substrate recognition. *Mol Cell*. 2007;27(2):324–331. https://doi.org/10.1016/j.molcel.2007.06.006.
228. Szczepinska T, Kalisiak K, Tomecki R, Labno A, Borowski LS. DIS3 shapes the RNA polymerase II transcriptome in humans by degrading a variety of unwanted transcripts. *Genome Res*. 2015;25(11):1622–1633. https://doi.org/10.1101/gr.189597.115.
229. Reimão-Pinto MM, Manzenreither RA, Burkard TR, et al. Molecular basis for cytoplasmic RNA surveillance by uridylation-triggered decay in Drosophila. *EMBO J*. 2016;35:2417–2434. https://doi.org/10.15252/embj.
230. Pashler AL, Towler BP, Jones CI, Newbury SF. The roles of the exoribonucleases DIS3L2 and XRN1 in human disease. *Biochem Soc Trans*. 2016;44(5):1377–1384. https://doi.org/10.1042/BST20160107.
231. Towler BP, Jones CI, Harper KL, Waldron JA, Newbury SF. A novel role for the 3′-5′ exoribonuclease Dis3L2 in controlling cell proliferation and tissue growth. *RNA Biol*. 2016;13(12):1286–1299. https://doi.org/10.1080/15476286.2016.1232238.
232. Basar E, Walch M, Lieberman J. Apoptosis triggers specific, rapid and global mRNA decay with 3′ Uridylated intermediates degraded by DIS3L2. *Cell Rep*. 2015;11(7):1079–1089. https://doi.org/10.1016/j.celrep.2015.04.026.
233. Chang H, Triboulet R, Thornton JE, Gregory RI, Program C, Cell HS. A role for the Perlman syndrome exonuclease Dis3l2 in the Lin28-let-7 pathway. *Nature*. 2013;497(7448):730–748. https://doi.org/10.1038/nature12119.
234. Haas G, Cetin S, Messmer M, et al. Identification of factors involved in target RNA-directed microRNA degradation. *Nucleic Acids Res*. 2016;44(6):2873–2887. https://doi.org/10.1093/nar/gkw040.
235. Razew M, Warkocki Z, Taube M, et al. Structural analysis of mtEXO mitochondrial RNA degradosome reveals tight coupling of nuclease and helicase components. *Nat Commun*. 2018;9(1):1–12. https://doi.org/10.1038/s41467-017-02570-5.
236. Dziembowski A, Malewicz M, Minczuk M, Golik P, Dmochowska A, Stepien PP. The yeast nuclear gene DSS1, which codes for a putative RNase II, is necessary for the function of the mitochondrial degradosome in processing and turnover of RNA. *Mol Gen Genet*. 1998;260(1):108–114. https://doi.org/10.1007/s004380050876.
237. Malecki M, Jedrzejczak R, Stepien PP, Golik P. In vitro reconstitution and characterization of the yeast mitochondrial degradosome complex unravels tight functional interdependence. *J Mol Biol*. 2007;372(1):23–36. https://doi.org/10.1016/j.jmb.2007.06.074.
238. Kumakura N, Otsuki H, Tsuzuki M, Takeda A, Watanabe Y. Arabidopsis AtRRP44A is the functional homolog of Rrp44/Dis3, an exosome component, is essential for viability and is required for RNA processing and degradation. *PLoS One*. 2013;8(11):1–14. https://doi.org/10.1371/journal.pone.0079219.
239. Chapman C, Tisa LS. Identification and characterization of Photorhabdus temperata mutants altered in hemolysis and virulence. *Can J Microbiol*. 2016;11:1–11. https://doi.org/10.1139/cjm-2016-0102.
240. Purusharth RI, Klein F, Sulthana S, et al. Exoribonuclease R interacts with endoribonuclease E and an RNA helicase in the psychrotrophic bacterium Pseudomonas syringae Lz4W. *J Biol Chem*. 2005;280(15):14572–14578. https://doi.org/10.1074/jbc.M413507200.

241. Reva ON, Weinel C, Weinel M, et al. Functional genomics of stress response in Pseudomonas putida KT2440. *J Bacteriol.* 2006;188(11):4079–4092. https://doi.org/10.1128/JB.00101-06.
242. Erova TE, Kosykh VG, Fadl AA, Sha J, Horneman AJ, Chopra AK. Cold shock exoribonuclease R (VacB) is involved in Aeromonas hydrophila pathogenesis. *J Bacteriol.* 2008;190(10):3467–3474. https://doi.org/10.1128/JB.00075-08.
243. Hutchison CA, Peterson SN, Gill SR, et al. Global transposon mutagenesis and a minimal Mycoplasma genome. *Science.* 1999;286(5447):2165–2169.
244. Haddad N, Matos RG, Pinto T, et al. The RNase R from Campylobacter jejuni has unique features and is involved in the first steps of infection. *J Biol Chem.* 2014;289(40):27814–27824. https://doi.org/10.1074/jbc.M114.561795.
245. Tobe T, Sasakawa C, Okada N, Honma Y, Yoshikawa M. vacB, a novel chromosomal gene required for expression of virulence genes on the large plasmid of Shigella flexneri. *J Bacteriol.* 1992;174(20):6359–6367. https://doi.org/10.1128/jb.174.20.6359-6367.1992.
246. Zhang Q, Siegel TN, Martins RM, et al. Exonuclease-mediated degradation of nascent RNA silences genes linked to severe malaria. *Nature.* 2014;513(7518):431–435. https://doi.org/10.1038/nature13468.
247. Rosenberg E, Ben-Shmuel A, Shalev O, Sinay R, Cowman A, Pollack Y. Differential, positional-dependent transcriptional response of antigenic variation (var) genes to biological stress in Plasmodium falciparum. *PLoS One.* 2009;4(9):1–8. https://doi.org/10.1371/journal.pone.0006991.
248. Zhang Q, Scherf A. RNase II: a new player enters the game. *RNA Biol.* 2015. https://doi.org/10.1080/15476286.2015.1038013.
249. Robinson SR, Viegas SC, Matos RG, et al. DIS3 isoforms vary in their endoribonuclease activity and are differentially expressed within haematological cancers. *Biochem J.* 2018. https://doi.org/10.1042/BCJ20170962.
250. Robinson SR, Oliver AW, Chevassut TJ, Newbury SF. The 3' to 5' exoribonuclease DIS3: from structure and mechanisms to biological functions and role in human disease. *Biomolecules.* 2015;5(3):1515–1539. https://doi.org/10.3390/biom5031515.
251. Weißbach S, Langer C, Puppe B, et al. The molecular spectrum and clinical impact of DIS3 mutations in multiple myeloma. *Br J Haematol.* 2015;169(1):57–70. https://doi.org/10.1111/bjh.13256.
252. Astuti D, Morris MR, Cooper WN, et al. Germline mutations in DIS3L2 cause the Perlman syndrome of overgrowth and Wilms tumor susceptibility. *Nat Genet.* 2012;44(3):277–284. https://doi.org/10.1038/ng.1071.

CHAPTER SIX

Different Methods of Delivering CRISPR/Cas9 Into Cells

Arun Pandian Chandrasekaran*, Minjung Song†, Kye-Seong Kim*,‡,1, Suresh Ramakrishna*,‡,1

*Graduate School of Biomedical Science and Engineering, Hanyang University, Seoul, South Korea
†Department of Food Biotechnology, College of Medical and Life Science, Silla University, Seoul, South Korea
‡College of Medicine, Hanyang University, Seoul, South Korea
[1]Corresponding authors: e-mail addresses: ks66kim@hanyang.ac.kr; suri28@hanyang.ac.kr; suresh.ramakris@gmail.com

Contents

1. Introduction 158
   1.1 Programmable Nucleases 158
   1.2 Genome Silencing vs Genome Editing 161
   1.3 Advantages of CRISPR Over ZFN and TALEN 161
2. Delivery Methods of CRISPR/Cas9 for Genome Editing 162
   2.1 Viral-Mediated Delivery 163
   2.2 Non-viral Vectors 167
   2.3 Physical Methods 168
3. Opportunities and Challenges in CRISPR/Cas9 Delivery to Stem Cells 169
4. Conclusions and Future Perspectives 170
Acknowledgments 171
References 171

Abstract

Clustered regularly interspaced short palindromic repeats (CRISPR)/CRISPR-associated system (Cas) is comprised of repetitive bases followed by short fragments of DNA from a previously invading organism that provide immunity to the most prokaryotic organisms. An RNA-dependent spacer is required for CRISPR/Cas9 to recognize the target DNA. Delivery of the CRISPR/Cas9-guide RNA (gRNA) complex to any cell results in modification of the target sequence. The CRISPR/Cas9-mediated genome editing technique is currently in the spotlight and has several research interests, including molecular medicine and agriculture. There are several factors that hinder the delivery of this complex, such as the large size of the plasmid or high dosage of the chemical agent. There are several methods available to deliver CRISPR/Cas9 and its components to the target cells. It includes viral, non-viral and physical methods to deliver plasmid or ribonucleoprotein (RNP) of CRISPR components. But in vivo CRISPR/Cas9 delivery remains challenging to the researchers due to insertional mutagenesis, targeted delivery, immunogenicity, and off-targets. However, studies suggesting that the CRISPR/Cas9-RNP delivery can

Progress in Molecular Biology and Translational Science, Volume 159
ISSN 1877-1173
https://doi.org/10.1016/bs.pmbts.2018.05.001

overcome these hurdles. Here, we review the various methods for delivery of CRISPR/Cas9 and gRNA to several cell lines, highlighting the limitations of each approach, and suggest possible alternative methods.

1. INTRODUCTION

The growth of genome editing technologies involving zinc finger nucleases (ZFNs), transcription activator-like effector nucleases (TALENs), or clustered regularly interspaced short palindromic repeats (CRISPR)/CRISPR-associated system (Cas) over a relatively short period has been rapid in many fields of research. The notion of genome editing by engineering nucleases has recently become popular. Site-specific alteration of the genome has transformed science and holds great promise for molecular medicine. Generally, genome editing indicates the insertion, deletion or replacement of a gene in the living organism using engineered nucleases. For the past two decades, researchers have used an array of proteins that recognize specific DNA sequences in the genome to efficiently edit the DNA using nucleases that result in DNA double-strand breaks (DSBs).[1] This activates DNA repair pathways including homologous recombination (HR) and non-homologous end joining.[2–4] Some of the important features of engineered nucleases are given in Table 1.

1.1 Programmable Nucleases

The first class of programmable nuclease was ZFNs, which can bind to specific target DNA through zinc finger proteins (ZFPs) (Fig. 1A). This 30-amino acid structure has a type II restriction endonuclease domain called FokI that is used to cleave the DNA. Due to the large number of ZFPs, the probability of off-target binding is very high and leads to undesired mutations.[17] A few years after the discovery of ZFNs, the search for a better alternative to ZFNs led to the introduction of TALENs. A typical TALEN involves fusion of a transcription activator-like effector (TALE) and a DNA cleavage domain (FokI) (Fig. 1B).[8] This complex introduces site-specific cleavage upon delivery into the cells. Initially, the wild-type FokI restriction enzyme was used to introduce the cleavage; however, in subsequent studies, FokI domain variants with mutations have also been employed to improve the cleavage activity[18] and specificity.[19,20] After much research on ZFNs and TALENs, the CRISPR/Cas9 system was introduced in the field of genome editing technology.

Table 1 Significant Features of Engineered Nucleases

			CRISPR Variants				
Type of Nuclease	**ZFN**	**TALEN**	**Cas9**	**Cpf1**	**C2c1**	**C2c2**	**Cas13b**
DNA binding partner	ZF proteins	TALE proteins	crRNA and tracrRNA	crRNA	crRNA and tracrRNA	crRNA	crRNA
Endonuclease	FokI	FokI	Cas9	Cpf1	C2c1[a]	C2c2[a]	Cas13b[a]
PAM	NA	NA	NGG	TTTN	TTN	PFS[b]	Flanking PFS[b]
Target length (nt)	18–36	30–40	20	23	20	28	30
Binding specificity ratio	1:3	1:1	1:1	1:1	1:1	1:1	1:1
Off-target effect	High	Low	Inconsistent	Low	Low	Low	Unknown
Reference	5–7	8	9–11	12	13	14,15	16

[a]RNA-guided RNA editing nuclease.
[b]Protospacer flanking sequence.

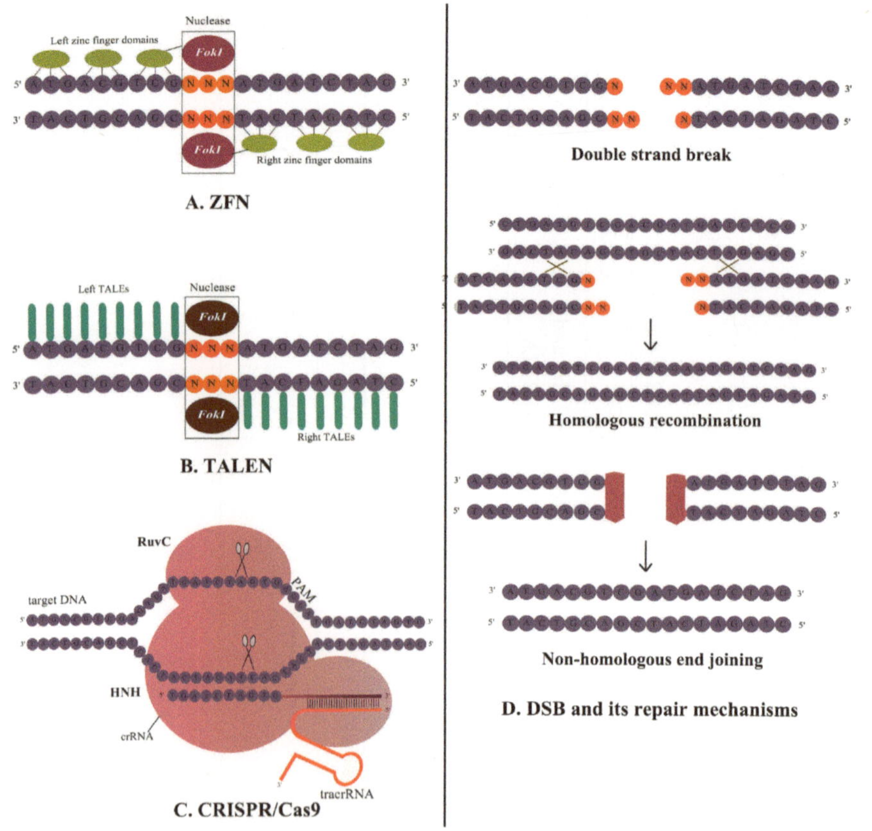

Fig. 1 Summary of genome editing tools. (A) Schematic representation of the ZFNs targeting the DNA molecule. Each individual right and left zinc finger domain targets each codon of target DNA. A FokI enzyme is attached to the both zinc fingers to mediate the site-specific cleavage. (B) Diagrammatic depiction of genome editing by TALENs. Similar to ZFNs, here the individual TALE protein recognizes a specific nucleotide of the target DNA and a FokI endonuclease is linked to TALE proteins for the cleavage. (C) Diagram showing the CRISPR/Cas9-mediated genome editing. The Cas9 nuclease has two domains, RuvC and HNH which is used to create the DSB. Here the Cas9 endonuclease is guided by crRNA and tracrRNA to cleave the target site. The crRNA and tracrRNA together called gRNA. (D) Illustration images of the DNA repair mechanisms after the site-specific cleavage by ZFNs, TALENs or CRISPR/Cas9.

Originally, the CRISPR system is an adaptive immunity of bacteria and archaea for protection against intrusive organisms.[21] Subsequently, CRISPR systems have been implemented in genome editing technology to target a specific locus. Three types and 10 subtypes of CRISPR/Cas have been suggested, and each type has core proteins and a definite module.[22] Among all available types, Cas9, which belongs to type II, is the most widely studied for genome editing.[23–25] CRISPR/Cas9 is easy to design and manipulate and efficient for genome editing in a wide variety of organisms.[9] Cas9 requires CRISPR RNA (crRNA) and trans-activating crRNA to direct the nuclease for sequence-specific cleavage. A protospacer adjacent motif (PAM) containing 2–5 nucleotides that are essential for targeting the DNA (Fig. 1C).[26]

1.2 Genome Silencing vs Genome Editing

The genome editing technique involves the complete eradication of gene expression and generation of DNA DSBs that result in "knockout" of a particular gene. In contrast, "knockdown" of the gene of interest by RNA interference (RNAi) or gene silencing methods is the most studied method among eukaryotes. The concept of RNAi was discovered in the worm *Caenorhabditis elegans*, a nematode in which the response to dsRNA results in target-specific gene silencing.[27] This novel research led to the concept that a number of distinct homology-dependent gene silencing events might share a general biological function. The mechanism behind this approach is the production of exogenous short interfering RNAs or short hairpin RNAs that are specific to the target gene. The resultant complex is handled by Dicer, affecting the downregulation of gene expression.

Recently, catalytically dead Cas9 (dCas9) was produced that does not contain endonuclease activity. dCas9 was used for gene silencing in mammalian cells via CRISPR interference (CRISPRi) and in *E. coli*.[28] The differentiation conditions of human pluripotent stem cells (hPSCs) can be regulated by dCas9 to study its pathway.[29,30] The main advantages of CRISPRi are the ability to target multiple genes and fewer off-target effects. Both knockdown and knockout approaches are essential and based on the research application.

1.3 Advantages of CRISPR Over ZFN and TALEN

The CRISPR/Cas9 system is a powerful technique that has been utilized in stem cells, knockout strategies,[31] generation of engineered animal models,[32] and cancer modeling.[33] The implications of using the CRISPR/Cas9 technique to edit the genome are relevant to several essential strategies including the development of antiviral therapies, restoration of unfavorable mutations, and model systems. Recently, the essential characteristics of mouse and human developmental lineage-restricted *BCL11A* gene were annotated using pooled CRISPR/Cas9–gRNA. As a result, erythroid enhancer of *BCL11A* was verified as a potential therapeutic target for the reinduction of fetal hemoglobin.[34]

The prerequisite for engineering of ZFNs and TALENs is the design of proteins that target DNA molecules, followed by nuclease assembly. One nuclease is used for right ZFN/TALEN and another for left ZFN/TALEN. In contrast, the CRISPR method of editing the genome is based on a single gRNA without the requirement for enzyme engineering. The Cas9 enzyme remains the same for all targeting sequences, and only the gRNAs must be customized. The highlight of the CRISPR/Cas9 system is the short gRNA sequence of approximately 20 base pairs targeting the specific region of a genome. Furthermore, gRNAs are easier and simpler to customize than other nucleases. Because of the versatility and specificity of the CRISPR/Cas9 system, laboratories with limited resources can utilize this technique for a wide variety of applications. Also, the short length of gRNA makes it more convenient to deliver to the target site than the high-MW ZFNs and TALENs. Several questions have arisen among the scientific community over the last few years, mainly regarding off-target activity and translation to the clinic, for example: Does the enzyme cleave random site(s) other than the target? What is the strategy for the CRISPR/Cas9 system to clear clinical trials? To overcome the off-target effects, whole-genome sequencing and deep sequencing strategies should be involved in the identification of off-targets induced by genome editing. Such analyses are important for human safety and to remove obstacles to clinical trials.

2. DELIVERY METHODS OF CRISPR/Cas9 FOR GENOME EDITING

Currently, there are several methods for delivering CRISPR systems in vivo and in vitro (Table 2). Both delivery vectors and physical methods are widely implemented for the efficient delivery of CRISPR/Cas9-mediated genome editing. Delivery vectors such as viral and non-viral vectors can accommodate mRNA or plasmid expressing the nucleases to target cells or tissues. Alternatively, physical methods including electroporation, laser, ballistic delivery, physical energy, or microinjection can be exploited for the delivery of nuclease into cells.[43] Basically, non-viral vectors are preferable because of the limitations of viral vectors such as carcinogenesis,[44] limited encapsulating capacity,[45] and immunogenicity.[46] Because of the low delivery efficiency in vivo, only a limited number of non-viral vectors for gene therapy have entered into clinical trials.[47,48] In this context, we summarized methods for delivery of CRISPR/Cas9 systems into cells (Fig. 2).

Table 2 Summary of CRISPR/Cas9 Delivery Systems for Various Applications in Stem Cells

Delivery Methods	Stem Cells	Target Gene(s)	Applications	Reference
Lentiviral vector	hPSCs	*NF1*, *MED12*, *CUL3*, *TADA1*, and *TADA2B*	Gene knockout	35
Adeno-associated viral vector	hHSCs	*HBB*	Gene correction	36
Electroporation	iPSCs	*HBB*	β-Thalassemia	37
		RPGR	Retinitis pigmentosa correction	38
	hPSCs	*SH2B3*	RBC production	39
		AKT2, *CELSR2*, *CIITA*, *GLUT4*, *LINC00116*, and *SORT1*	Comparative study between TALEN and CRISPR	40
Nucleofection	iPSCs	*CFTR*	Cystic fibrosis	41
	hHSCs	*B2M* and *CCR5*	Gene knockout	42

2.1 Viral-Mediated Delivery

Viral-mediated delivery is accomplished through two mechanisms: infection and replication. During the infection stage, a virus can recognize and enter a specific cell, and the viral genome will be released into the nucleus (in case of DNA) or cytoplasm (in case of RNA) for replication. After replication of the viral genome in the cells, reproduced virions exit the cells. The infection stage starts again in neighboring cells, and the infection–replication cycle continues.[49] The virus containing the delivery material (programmed genome editing nuclease) is transported to the target cells and gene therapy can be achieved by genome editing. A number of viral vectors have been developed such as adenoviral vectors (AdVs), adeno-associated viral vectors (AAVVs), and lentiviral vectors (LVs).

2.1.1 Adeno-Associated Viral Vector-Mediated Delivery

Delivery of Cas9 into different cell lines using an AAVV system has been reported.[50–52] A suitable guide RNA (gRNA) and SaCas9

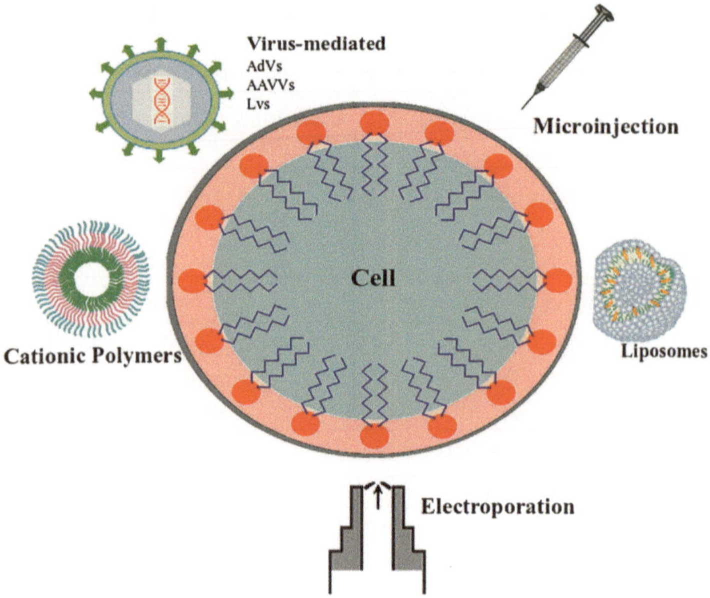

Fig. 2 Schematic representations of the different delivery methods for the successful delivery of CRISPR/Cas9 and gRNA plasmids to the cells.

(Cas9 from *Streptococcus aureus*) nuclease were successfully delivered to the livers of adult mice using the AAVV-mediated delivery system,[53] leading to successful disruption of the *Pcsk9* gene in mouse liver (40% disruption). Furthermore, SpCas9 (Cas9 from *Streptococcus pyogenes*) was also delivered and showed successful mutation of the *Pcsk9* gene with a rate higher than 50%.[54] Knockout of the *Pcsk9* gene led to a low cholesterol level in the mice, as PCSK9 protein is involved in LDL receptor degradation. In another study, delivery of CRISPR/Cas9 through AAVVs was achieved with restoration of functional deficiency of mouse muscle.[55] The mRNA or cDNA of SaCas9 was encapsulated in either AAVV8 or AAVV9 for delivery into tibialis anterior muscle of *mdx* mice for correction of Duchenne muscular dystrophy (DMD).[55–57] The mouse model *mdx* has a mutation in exon 23 which is responsible for DMD. An AAVV-mediated delivery method was proposed to treat DMD in the mouse model by employing CRISPR/Cas9-based editing technique. SaCas9 was used for in vivo genome editing. On the other hand, AAVV expressing sgRNAs were also prepared to target introns 22 and 23 to create deletion. These introns flanking to the exon 23 on both the sides and thus the deletion of the two introns resulted in the complete elimination of exon 23. Equal volume of Cas9 and sgRNA were injected into the muscle of *mdx* mice and harvested after 8 weeks for analysis. End-point PCR and droplet digital PCR showed the expected deleted region (approx. 1171 bp) and approximately 2% of all alleles showed exon 23 deletion. Sanger and deep sequencing results further confirmed the deletion of exon 23. The results showed that muscle force was enriched and muscle biochemistry was strongly improved. This research demonstrated the potential of CRISPR/Cas9 delivery by AAVV as a prospective therapy to treat DMD.

The AAVV delivery system is commonly used for gene therapy due to its long-lasting transgene expression, mild immune response, high infection efficiency, and general safety and is now being tested in clinical trials.[58] However, the limited packaging range of approximately 4.7 kb makes it inappropriate for larger Cas9 variants.[26,59] To overcome this limitation, several distinct solutions have been proposed. Smaller orthologs of Cas9 (SaCas9) with a size of 3.3 kb have been bundled into a single AAVV,[53] and another ortholog called *Streptococcus thermophilus* Cas9 (St1Cas9), which was also 3.3 kb, was similarly packed into a single AAVV.[60,61] Furthermore, other studies employed the dual-AAVV system in which one AAVV delivers SpCas9 and another delivers gRNA.[62,63] The disadvantage of this system is the reduced probability of delivering both viral vectors to the same cell, which decreases the efficiency of delivery.

2.1.2 Lentiviral Vector-Mediated Delivery

The simultaneous delivery of both Cas9 and gRNA into the host cells can be accomplished by lentiviral vectors. Here, a cell line expressing Cas9 is not a prerequisite for the co-delivery of Cas9 and gRNA.[35] Basically, LVs are initially acclimatized from HIV-1, which yields an extremely effective viral vector due to its tropism, capability to infect post-mitotic cells, and larger cargo capacity of 9.7 kb.[64] Ex vivo gene therapy can also be achieved using LVs, particularly in T cells (T lymphocytes) and hematopoietic stem cells (HSCs). In a recent study, LVs were exploited to treat β-thalassemia and Wiskott–Aldrich syndrome and for T-lymphocyte immunotherapy.[65] Use of integrase-deficient lentivirus vectors that have viral LTRs and integrase mutations significantly decreases the risk of insertional mutagenesis, but does not entirely eradicate the risk.[66–68] Currently, various research groups are utilizing LV-mediated delivery of gRNA and Cas9 to create knockout libraries.[35,69] These LVs can incorporate into the target genome without changing its original form and express both gRNA and Cas9.

LVs have great potential for further applications in in vivo gene therapy. Proof of concept of in vivo therapy using an LV-dependent system for the delivery of CRISPR–Cas9 system is now being developed. Recently, a pre-packed LV-Cas9 protein system have been developed to edit CCR5 gene in TZM-bl cells.[70] Also, they designed a specific sgRNA targeting the long terminal repeat of HIV to disrupt HIV proviral DNA in the J-LAT cells. The CCR5 sgRNA displayed minimal off-target on chromosome 4. The above evidence has further confirmed the efficiency of LV-mediated delivery of Cas9 for safe access for gene therapy applications. In another study, dual LV system was developed to get effective gene knockout in difficult to transfect human and mouse cell lines and primary cell lines.[71] This dual vector consists of an inducible sgRNA cassette and a constitutive expression Cas9 vector linked with mCherry. With the help of this system, they induced mutation in Trp53 in primary HSPCs and resulted in the growth of lymphoma that is driven by MYC. Thus, the combined action of dual LV and CRISPR/Cas9 systems can be utilized for annotating novel mutations in tumor suppressor genes and oncogenes that can accelerate tumor development.[71] However, further analysis might require that the delivery of the substrates and effector proteins is well balanced and timed. Nuclease-deficient Cas9 (dCas9) delivery can be employed, or non-integrating LVs can be used for nuclease applications. Also, problems concerning the integration of Cas9 into the target genome using LVs are not completely resolved.[72]

2.1.3 Adenovirus-Mediated Delivery

Adenoviridae is a family of adenoviruses that were isolated from human adenoid tissues in 1953. Generally, the concept of AdV-mediated delivery can be used for in vivo and in vitro applications.[73] Delivery of RNA-guided nuclease into several cell lines was successfully accomplished.[74] The researchers investigated second-generation fiber-modified AdVs expressing gRNA or the Cas9 component transduced into a recombinant allele or the safe harbor gene called *AAVS1* that can be generated to high titers.[74] Although AdVs generate immune responses when transduced into hepatocytes, genome editing of somatic cells by AdVs can generate prolonged liver phenotypes.[75] The SpCas9 system was delivered by AdV to target the *Pten* gene of mouse liver, and the genome editing efficiencies of AdVs and hydrodynamic injection-mediated delivery were compared. After 2 weeks of treatment, the hydrodynamic injection-mediated delivery was less efficient than AdVs. In addition, steatohepatitis and hepatomegaly were observed after 4 weeks of AdV genome editing treatment.[75]

2.2 Non-viral Vectors

Several non-viral vector systems have been developed and successfully employed for safe delivery of CRISPR/Cas9 to cells. Noteworthy delivery methods include polymeric materials, liposomes, cell-penetrating peptides (CPPs), and cationic nanocarriers. The main advantages of non-viral vectors are the capability to accommodate components of a large size for delivery, their reduced or non-hazardous nature, and easy generation. For these reasons, many scientists select this type of system for the delivery of nucleases.

2.2.1 Cationic Vectors

There are two broad categories of cationic based vectors: cationic polymer-based and cationic lipid-based vectors. The probability of uptake of the delivery components (Cas9 or gRNA plasmid) is high because of their anionic nature. Thus, by nature, the cationic vectors are more suitable and efficient for delivery of anionic gRNA or Cas9 to the cells. Effective delivery of Cas9 was achieved using cationic polymer nanoparticles (NPs).[76] Polyethyleneimine (PEI) is a common cationic polymer that is used for the delivery of CRISPR components. The molecular weight (MW), the amount of branched or linear structure, and structural characteristics are significant properties that determine the transfection efficiency and toxicity to the cells. Also, the secondary amine present in the PEI helps to retain the DNA in the cells.[76] The stability of PEI is based on its high MW, high number of branched structures, and high cationic charge; however, it could cause toxicity to cells. PEI-based delivery of Cas9 for in vivo targeting of tumor suppressor genes (*Pten*, *Trp53*, *Nf1*) in mouse brain successfully caused gene disruption.[77]

Cationic lipid-based vectors are also useful for the efficient delivery of Cas9 and gRNA. For instance, gRNA and Cas9 ribonucleoprotein (RNP) complexes were successfully delivered to mouse hair cells by cationic liposomes, and 80% gene modification was achieved.[78] The vectors were injected into the inner ear rather than using the intravenous route, as injection is efficient and similar to the clinical technique. Another technique for the delivery of CRISPR/Cas9 is by employing 7C1 NPs, which can be produced by mixing low-MW PEI with C15 epoxide-terminated lipids. This conglomeration method enhanced the mutation efficiency in the cardiovascular and pulmonary endothelium and generated Cre-based Cas9 mice.[51]

2.2.2 Cell-Penetrating Peptides

Technical limitations such as susceptibility to enzymatic degradation, anionic nature, and high MW make DNA delivery into target cells very difficult. CPPs have several advantages and are proven to be successful in delivering therapeutics against multiple diseases.[79] CPPs carrying Cas9–gRNA have been used for successful genome alteration in various cell types (HeLa cells, dermal fibroblasts, human embryonic stem cells (hESCs), HEK293T cells, and embryonic carcinoma cells) with low off-target activity. In one study, CPP was conjugated with gRNAs and recombinant Cas9 proteins to cause gene disruptions at the target site in human cells lines. This approach has some advantages over other non-viral methods because of its chemical reagent-free delivery and fewer off-target effects.[80]

2.2.3 Other Non-viral Methods

Several studies have involved CRISPR/Cas9 delivery using cationic NPs. Recently, CRISPR/Cas9 was delivered by bioreducible lipid NPs with 70% genome efficiency.[81] A hydrodynamic-based injection for the delivery of gRNA–Cas9 complex to target cells was recently reported.[82] Hydrodynamic injections are high-volume injections that are delivered at high speed into the vasculature for efficient delivery of Cas9 and gRNA plasmids and have been used in many in vivo analyses targeting the liver.[82,83] In this study, a cancer model had been generated using CRISPR/Cas9 in mice. A plasmid co-expressing Cas9 and sgRNA was injected to the liver by the hydrodynamic injection. The plasmids target the tumor suppressor genes *p53* and *Pten*. The DNA sequencing results confirmed the biallelic mutations happened in both *p53* and *Pten*.[83] This data exhibits the possibility of direct mutation of tumor suppressor genes in the liver using CRISPR/Cas9 that explains the progress in liver cancer models and genomics. However, this method is not extensively used because of possible damage to the heart and liver.[84]

2.3 Physical Methods

There are several physical methods for delivery of Cas9 nuclease into cells including microinjection and electroporation. Because of its target specificity, high reproducibility, and simplicity, this approach is a promising technology for gene therapy without the limitations associated with viral vectors. Also, these methods can be readily applied to different cell lines in vivo and in vitro.[85] For direct delivery of the constructs to a target cell, a needle with diameter ranging from 0.5 to 20 μM is generally inserted into the zebrafish[86] and mosquito embryos.[87] The overall experiment is aided by a special microscope equipped with a micromanipulator.[86–88] Alternatively, electroporation generates pores on the cells and thus enables the plasmids to enter the cells. Usually, the electroporator is used to produce electric pulses to create pores.[85,89–91] Efficient knockout of the *SH2B3* gene in hESCs was accomplished by co-transfection of Cas9 and gRNA plasmids. In this knockout analysis, 25 μg of each plasmid was electroporated into hESCs. The resultant suppression of the *SH2B3* gene in hESCs enhanced erythroid expansion and differentiation. Similarly, *SH2B3* gene suppression in both HSCs and pluripotent stem cells (PSCs) resulted in erythroid expansion.[39] CRISPR/Cas9-based targeting of the cystic fibrosis (CF) gene (*CTFR*) in patient-derived induced pluripotent stem cells (iPSCs) significantly corrected the phenotype. The CRISPR/Cas9 plasmid, gRNA for *CTFR*, GFP, and donor template were co-nucleofected into iPSCs from CF patients. Consequently, the corrected iPSCs differentiated into epithelial cells, and their function was analyzed.[41]

3. OPPORTUNITIES AND CHALLENGES IN CRISPR/Cas9 DELIVERY TO STEM CELLS

The improvement of target-specific nucleases as research tools parallels recent developments in iPSCs. Reconsidering the Cas9 technique as an engineered nuclease eliminates the barriers that restricted the prospects of genome editing by offering a common platform. The main advantage of this system over ZFNs and TALENs is that the DNA-targeting specificity is determined by the gRNA and does not involve the arduous process of designing DNA binding proteins. Thus, integrating the cellular adaptability of iPSC differentiation with straightforward genome editing by the CRISPR/Cas9 system is a powerful technique and has become a benchmark tool in stem cell research and disease modeling.[92,93]

An inducible cassette expressing Cas9 with gRNA was delivered to the *AAVS1* locus to enable the generation of isogenic hPSCs. Doxycycline was used to induce the expression of Cas9.[94] Here, iCRISPR (combination of TALEN and CRISPR/Cas9) was developed for the efficient biallelic knockout of hPSCs and for knock-in specific alternations in the nucleotide to generate disease models in hPSCs. First, AAVS1-TALEN along with two donor constructs were electroporated to the first intron of the AAVS1 in hESCs and hiPSCs cell lines. A high biallelic mutation was observed in >50% of clones which was confirmed by Southern blot and qRT-PCR revealed the Cas9 expression after doxycycline induction to all iCas9 clones. Furthermore, the pluripotency markers (NANOG, OCT2, and SOX2) were expressed in iCas9 hPSCs and retained its stemness property. Altogether, the iCRISPR can be used to study the pleiotropy in human disease and to examine the complex genetic interactions. Also, it has the ability to assist the high-throughput analysis in hPSCs. A loss-of-function study of *EZH2* was performed, and haploinsufficiency for *EZH2* was revealed in hematopoietic differentiation.[95] Recently, effective delivery of CRISPR/Cas9 into human hematopoietic stem and progenitor cells (HSPCs) was accomplished by electroporation with chemically modified gRNA targeting the cells.[96] The chemical modifications include 2′-*O*-methyl, 2′-*O*-methyl 3′ phosphorothioate, or 2′-*O*-methyl 3′ thioPACE were integrated at the 5′ and 3′ ends.[96] Another recently published study described targeting the *HBB* gene in HSPCs.[36] HR-dependent genome editing in $CD34^+$ HSPCs from peripheral blood was performed using CRISPR/Cas9 ribonucleoprotein paired with recombinant AAVV serotype 6 (rAAVV6) donor. To attain efficient gene editing, the chemically modified sgRNAs were co-transfected with Cas9 RNP or mRNA. The *HBB*-single-stranded AAVV6 (ssAAVV6) was tagged with GFP to check the transfection efficiency. Both RNP and mRNA resulted in high INDEL when HSPCs was electroporated. After the electroporation of Cas9 RNP, *HBB*-ssAAVV6 donor was introduced and 29% of stable GFP expression was obtained. On the other hand, mRNA system showed lower efficiency (15%). Also, the off-target and cytotoxicity were low for RNP platform. The above data revealed that CRISPR/Cas9-based editing of the *HBB* gene in HSCs has the potential for translation to clinical medicine.

4. CONCLUSIONS AND FUTURE PERSPECTIVES

Both non-viral and viral vectors have been implemented for the delivery of CRISPR/Cas9 for various applications. Non-viral vectors for the delivery of CRISPR/Cas9 in the form of plasmid, mRNA or protein along with gRNA has appeared as a potential delivery approach with intrinsic strength and ability. On the other hand, Cas9 delivery approaches by viral vectors have attained some level of in vivo therapeutic gene editing but effective editing remains a challenge. The effective delivery of CRISPR components to the target is the major challenge with respect to specificity and efficacy. Depending on the nuclease expression and modes of delivery, both immune response and off-targets are possible. Both non-viral and viral methods are being assessed for the delivery of CRISPR/Cas9 to cells in vivo or ex vivo. On the basis of the evidences, the in vivo delivery of Cas9 via PEL, CPP, AAV, and cationic liposomes are successful. Furthermore, hydrodynamic injection has been applied to carry a large size of CRISPR/Cas9 plasmid into mouse hepatic tissue.[82,83] On the other hand, physical methods include microinjection and electroporation have succeeded in the uptake of plasmids to the target. Additionally, a formulation consist of cationic surfactant benzalkonium chloride and non-ionic poloxamer CRL 1005 has gone into clinical trial phase II/III.[97]

The non-viral delivery of Cas9 plasmid or protein decreases off-targets and immune response in vivo. Also, these vectors have lesser advantage over viral vectors in various conditions specifically gene knock-in. However, constant expression of CRISPR/Cas9 using viral vectors induces off-targets, immune response and thus must be improved. In the near future, donor template and sgRNA may be prepared into viral vectors for constant expression and transient CRISPR/Cas9 can be introduced via non-viral and administered for efficient DNA cleavage. Thus, the blending of multi-administered non-viral vectors with viral vector will be the most favorable method for molecular medicine based on CRISPR/Cas9 technology.

ACKNOWLEDGMENTS

We would like to thank all of Suri's laboratory members for their helpful discussion. This study was supported by a grant from the National Research Foundation of Korea (2018M3A9H3022412, 2017M3A9C6061361, 2017M3A9E4048172, 2017R1A2B2008727).

REFERENCES

1. Carroll D. Genome engineering with targetable nucleases. *Annu Rev Biochem*. 2014;83: 409–439. https://doi.org/10.1146/annurev-biochem-060713-035418.
2. Smih F, Rouet P, Romanienko PJ, Jasin M. Double-strand breaks at the target locus stimulate gene targeting in embryonic stem cells. *Nucleic Acids Res*. 1995;23(24): 5012–5019. doi:5a0556 [pii].
3. Wyman C, Kanaar R. DNA double-strand break repair: all's well that ends well. *Annu Rev Genet*. 2006;40(1):363–383. https://doi.org/10.1146/annurev.genet.40.110405.090451.
4. Rouet P, Smih F, Jasin M. Expression of a site-specific endonuclease stimulates homologous recombination in mammalian cells. *Proc Natl Acad Sci USA*. 1994;91(13): 6064–6068. https://doi.org/10.1073/pnas.91.13.6064.
5. Moore M, Klug A, Choo Y. Improved DNA binding specificity from polyzinc finger peptides by using strings of two-finger units. *Proc Natl Acad Sci USA*. 2001;98(4): 1437–1441. https://doi.org/10.1073/pnas.98.4.1437.
6. Kim J-S, Lee HJ, Carroll D. Genome editing with modularly assembled zinc-finger nucleases. *Nat Methods*. 2010;7(2):91. author reply 91–92. https://doi.org/10.1038/nmeth0210-91a.

7. Ramirez CL, Foley JE, Wright DA, et al. Unexpected failure rates for modular assembly of engineered zinc fingers. *Nat Methods*. 2008;5(5):374–375. https://doi.org/10.1038/nmeth0508-374.
8. Boch J. TALEs of genome targeting. *Nat Biotechnol*. 2011;29(2):135–136. https://doi.org/10.1038/nbt.1767.
9. Ran FA, Hsu PD, Wright J, Agarwala V, Scott DA, Zhang F. Genome engineering using the CRISPR-Cas9 system. *Nat Protoc*. 2013;8(11):2281–2308. https://doi.org/10.1038/nprot.2013.143.
10. Cong L, Ran FA, Cox D, et al. Multiplex genome engineering using CRISPR/Cas systems. *Science*. 2013;339(6121):819–823. https://doi.org/10.1126/science.1231143.
11. Hsu PD, Scott DA, Weinstein JA, et al. DNA targeting specificity of RNA-guided Cas9 nucleases. *Nat Biotechnol*. 2013;31(9):827–832. https://doi.org/10.1038/nbt.2647.
12. Zetsche B, Gootenberg JS, Abudayyeh OO, et al. Cpf1 is a single RNA-guided endonuclease of a class 2 CRISPR-Cas system. *Cell*. 2015;163(3):759–771. https://doi.org/10.1016/j.cell.2015.09.038.
13. Yang H, Gao P, Rajashankar KR, Patel DJ. PAM-dependent target DNA recognition and cleavage by C2c1 CRISPR-Cas endonuclease. *Cell*. 2016;167(7):1814–1828:e12. https://doi.org/10.1016/j.cell.2016.11.053.
14. Weintraub K. CRISPR gene-editing system unleashed on RNA. *Nature*. 2016. https://doi.org/10.1038/nature.2016.20030.
15. Abudayyeh OO, Gootenberg JS, Konermann S, et al. C2c2 is a single-component programmable RNA-guided RNA-targeting CRISPR effector. *Science*. 2016;353(6299): aaf5573–aaf5573. https://doi.org/10.1126/science.aaf5573.
16. Smargon AA, Cox DBT, Pyzocha NK, et al. Cas13b is a type VI-B CRISPR-associated RNA-guided RNase differentially regulated by accessory proteins Csx27 and Csx28. *Mol Cell*. 2017;65(4):618–630:e7. https://doi.org/10.1016/j.molcel.2016.12.023.
17. Carroll D. Genome engineering with zinc-finger nucleases. *Genetics*. 2011;188(4): 773–782. https://doi.org/10.1534/genetics.111.131433.
18. Guo J, Gaj T, Barbas CF. Directed evolution of an enhanced and highly efficient FokI cleavage domain for zinc finger nucleases. *J Mol Biol*. 2010;400(1):96–107. https://doi.org/10.1016/j.jmb.2010.04.060.
19. Szczepek M, Brondani V, Büchel J, Serrano L, Segal DJ, Cathomen T. Structure-based redesign of the dimerization interface reduces the toxicity of zinc-finger nucleases. *Nat Biotechnol*. 2007;25(7):786–793. https://doi.org/10.1038/nbt1317.
20. Doyon Y, Vo TD, Mendel MC, et al. Enhancing zinc-finger-nuclease activity with improved obligate heterodimeric architectures. *Nat Methods*. 2011;8(1):74–79. https://doi.org/10.1038/nmeth.1539.
21. Jinek M, Chylinski K, Fonfara I, Hauer M, Doudna JA, Charpentier E. A programmable dual-RNA-guided DNA endonuclease in adaptive bacterial immunity. *Science*. 2012;337(6096):816–821. https://doi.org/10.1126/science.1225829.
22. Makarova KS, Koonin EV. Annotation and classification of CRISPR-Cas systems. *Methods Mol Biol*. 2015;1311:47–75. https://doi.org/10.1007/978-1-4939-2687-9_4.
23. Brouns SJJ, Jore MM, Lundgren M, et al. Small CRISPR RNAs guide antiviral defense in prokaryotes. *Science*. 2008;321(5891):960–964. https://doi.org/10.1126/science.1159689.
24. Barrangou R, Fremaux C, Deveau H, et al. CRISPR provides acquired resistance against viruses in prokaryotes. *Science*. 2007;315(5819):1709–1712. https://doi.org/10.1126/science.1138140.
25. Haft DH, Selengut J, Mongodin EF, Nelson KE. A guild of 45 CRISPR-associated (Cas) protein families and multiple CRISPR/Cas subtypes exist in prokaryotic genomes. *PLoS Comput Biol*. 2005;1(6):e60. https://doi.org/10.1371/journal.pcbi.0010060.
26. Mali P, Esvelt KM, Church GM. Cas9 as a versatile tool for engineering biology. *Nat Methods*. 2013;10(10):957–963. https://doi.org/10.1038/nmeth.2649.

27. Fire A, Xu S, Montgomery MK, Kostas SA, Driver SE, Mello CC. Potent and specific genetic interference by double-stranded RNA in Caenorhabditis elegans. *Nature.* 1998;391(6669):806–811. https://doi.org/10.1038/35888.
28. Qi LS, Larson MH, Gilbert LA, et al. Repurposing CRISPR as an RNA-guided platform for sequence-specific control of gene expression. *Cell.* 2013;152(5):1173–1183. https://doi.org/10.1016/j.cell.2013.02.022.
29. Wang T, Wei JJ, Sabatini DM, Lander ES. Genetic screens in human cells using the CRISPR-Cas9 system. *Science.* 2014;343(6166):80–84. https://doi.org/10.1126/science.1246981.
30. Kearns NA, Genga RMJ, Enuameh MS, Garber M, Wolfe SA, Maehr R. Cas9 effector-mediated regulation of transcription and differentiation in human pluripotent stem cells. *Development.* 2014;141(1):219–223. https://doi.org/10.1242/dev.103341.
31. Dow LE, Fisher J, O'Rourke KP, et al. Inducible in vivo genome editing with CRISPR-Cas9. *Nat Biotechnol.* 2015;33(4):390–394. https://doi.org/10.1038/nbt.3155.
32. Kraft K, Geuer S, Will AJ, et al. Deletions, inversions, duplications: engineering of structural variants using CRISPR/Cas in mice. *Cell Rep.* 2015;10(5):833–839. https://doi.org/10.1016/j.celrep.2015.01.016.
33. Maeder ML, Gersbach CA. Genome-editing technologies for gene and cell therapy. *Mol Ther.* 2016;24(3):430–446. https://doi.org/10.1038/mt.2016.10.
34. Canver MC, Smith EC, Sher F, et al. BCL11A enhancer dissection by Cas9-mediated in situ saturating mutagenesis. *Nature.* 2015;527(7577):192–197. https://doi.org/10.1038/nature15521.
35. Shalem O, Sanjana NE, Hartenian E, et al. Genome-scale CRISPR-Cas9 knockout screening in human cells. *Science.* 2014;343(6166):84–87. https://doi.org/10.1126/science.1247005.
36. Dever DP, Bak RO, Reinisch A, et al. CRISPR/Cas9 β-globin gene targeting in human haematopoietic stem cells. *Nature.* 2016;539(7629):384–389. https://doi.org/10.1038/nature20134.
37. Xie F, Ye L, Chang JC, et al. Seamless gene correction of β-thalassemia mutations in patient-specific iPSCs using CRISPR/Cas9 and piggyBac. *Genome Res.* 2014;24(9):1526–1533. https://doi.org/10.1101/gr.173427.114.
38. Bassuk AG, Zheng A, Li Y, Tsang SH, Mahajan VB. Precision medicine: genetic repair of retinitis pigmentosa in patient-derived stem cells. *Sci Rep.* 2016;6(1):19969. https://doi.org/10.1038/srep19969.
39. Giani FC, Fiorini C, Wakabayashi A, et al. Targeted application of human genetic variation can improve red blood cell production from stem cells. *Cell Stem Cell.* 2016;18(1):73–78. https://doi.org/10.1016/j.stem.2015.09.015.
40. Ding Q, Regan SN, Xia Y, Oostrom LA, Cowan CA, Musunuru K. Enhanced efficiency of human pluripotent stem cell genome editing through replacing TALENs with CRISPRs. *Cell Stem Cell.* 2013;12(4):393–394. https://doi.org/10.1016/j.stem.2013.03.006.
41. Firth AL, Menon T, Parker GS, et al. Functional gene correction for cystic fibrosis in lung epithelial cells generated from patient iPSCs. *Cell Rep.* 2015;12(9):1385–1390. https://doi.org/10.1016/j.celrep.2015.07.062.
42. Mandal PK, Ferreira LMR, Collins R, et al. Efficient ablation of genes in human hematopoietic stem and effector cells using CRISPR/Cas9. *Cell Stem Cell.* 2014;15(5):643–652. https://doi.org/10.1016/j.stem.2014.10.004.
43. Wells DJ. Gene therapy progress and prospects: electroporation and other physical methods. *Gene Ther.* 2004;11(18):1363–1369. https://doi.org/10.1038/sj.gt.3302337.
44. Baum C, Kustikova O, Modlich U, Li Z, Fehse B. Mutagenesis and oncogenesis by chromosomal insertion of gene transfer vectors. *Hum Gene Ther.* 2006;17(3):253–263. https://doi.org/10.1089/hum.2006.17.253.

45. Thomas CE, Ehrhardt A, Kay MA. Progress and problems with the use of viral vectors for gene therapy. *Nat Rev Genet*. 2003;4(5):346–358. https://doi.org/10.1038/nrg1066.
46. Bessis N, GarciaCozar FJ, Boissier M-C. Immune responses to gene therapy vectors: influence on vector function and effector mechanisms. *Gene Ther*. 2004;11(suppl 1): S10–S17. https://doi.org/10.1038/sj.gt.3302364.
47. Vectors Used in Gene Therapy Clinical Trials. 2002. http://www.wiley.com//legacy/wileychi/genmed/clinical/. Accessed July 20, 2002.
48. Yin H, Kanasty RL, Eltoukhy AA, Vegas AJ, Dorkin JR, Anderson DG. Non-viral vectors for gene-based therapy. *Nat Rev Genet*. 2014;15(8):541–555. https://doi.org/10.1038/nrg3763.
49. Vannucci L, Lai M, Chiuppesi F, Ceccherini-Nelli L, Pistello M. Viral vectors: a look back and ahead on gene transfer technology. *New Microbiol*. 2013;36(1):1–22.
50. Lai Y, Yue Y, Liu M, et al. Efficient in vivo gene expression by trans-splicing adeno-associated viral vectors. *Nat Biotechnol*. 2005;23(11):1435–1439. https://doi.org/10.1038/nbt1153.
51. Platt RJ, Chen S, Zhou Y, et al. CRISPR-Cas9 knockin mice for genome editing and cancer modeling. *Cell*. 2014;159(2):440–455. https://doi.org/10.1016/j.cell.2014.09.014.
52. Swiech L, Heidenreich M, Banerjee A, et al. In vivo interrogation of gene function in the mammalian brain using CRISPR-Cas9. *Nat Biotechnol*. 2015;33(1):102–106. https://doi.org/10.1038/nbt.3055.
53. Ran FA, Cong L, Yan WX, et al. In vivo genome editing using Staphylococcus aureus Cas9. *Nature*. 2015;520(7546):186–191. https://doi.org/10.1038/nature14299.
54. Ding Q, Strong A, Patel KM, et al. Permanent alteration of PCSK9 with in vivo CRISPR-Cas9 genome editing. *Circ Res*. 2014;115(5):488–492. https://doi.org/10.1161/CIRCRESAHA.115.304351.
55. Tabebordbar M, Zhu K, Cheng JKW, et al. In vivo gene editing in dystrophic mouse muscle and muscle stem cells. *Science*. 2016;351(6271):407–411. https://doi.org/10.1126/science.aad5177.
56. Nelson CE, Hakim CH, Ousterout DG, et al. In vivo genome editing improves muscle function in a mouse model of Duchenne muscular dystrophy. *Science*. 2016;351(6271):403–407. https://doi.org/10.1126/science.aad5143.
57. Long C, Amoasii L, Mireault AA, et al. Postnatal genome editing partially restores dystrophin expression in a mouse model of muscular dystrophy. *Science*. 2016;351(6271): 400–403. https://doi.org/10.1126/science.aad5725.
58. Mingozzi F, High KA. Therapeutic in vivo gene transfer for genetic disease using AAV: progress and challenges. *Nat Rev Genet*. 2011;12(5):341–355. https://doi.org/10.1038/nrg2988.
59. Hsu PD, Lander ES, Zhang F. Development and applications of CRISPR-Cas9 for genome engineering. *Cell*. 2014;157(6):1262–1278. https://doi.org/10.1016/j.cell.2014.05.010.
60. Horvath P, Romero DA, Coûté-Monvoisin A-C, et al. Diversity, activity, and evolution of CRISPR loci in Streptococcus thermophilus. *J Bacteriol*. 2008;190(4):1401–1412. https://doi.org/10.1128/JB.01415-07.
61. Deveau H, Barrangou R, Garneau JE, et al. Phage response to CRISPR-encoded resistance in Streptococcus thermophilus. *J Bacteriol*. 2008;190(4):1390–1400. https://doi.org/10.1128/JB.01412-07.
62. Hung SSC, Chrysostomou V, Li F, et al. AAV-mediated CRISPR/Cas gene editing of retinal cells in vivo. *Invest Ophthalmol Vis Sci*. 2016;57(7):3470–3476. https://doi.org/10.1167/iovs.16-19316.
63. Yang Y, Wang L, Bell P, et al. A dual AAV system enables the Cas9-mediated correction of a metabolic liver disease in newborn mice. *Nat Biotechnol*. 2016;34(3):334–338. https://doi.org/10.1038/nbt.3469.

64. Naldini L, Blömer U, Gallay P, et al. In vivo gene delivery and stable transduction of nondividing cells by a lentiviral vector. *Science*. 1996;272(5259):263–267.
65. Naldini L. Gene therapy returns to centre stage. *Nature*. 2015;526(7573):351–360. https://doi.org/10.1038/nature15818.
66. Philippe S, Sarkis C, Barkats M, et al. Lentiviral vectors with a defective integrase allow efficient and sustained transgene expression in vitro and in vivo. *Proc Natl Acad Sci USA*. 2006;103(47):17684–17689. https://doi.org/10.1073/pnas.0606197103.
67. Miyoshi H, Blömer U, Takahashi M, Gage FH, Verma IM. Development of a self-inactivating lentivirus vector. *J Virol*. 1998;72(10):8150–8157.
68. Wanisch K, Yáñez-Muñoz RJ. Integration-deficient lentiviral vectors: a slow coming of age. *Mol Ther*. 2009;17(8):1316–1332. https://doi.org/10.1038/mt.2009.122.
69. Sanjana NE, Shalem O, Zhang F. Improved vectors and genome-wide libraries for CRISPR screening. *Nat Methods*. 2014;11(8):783–784. https://doi.org/10.1038/nmeth.3047.
70. Choi JG, Dang Y, Abraham S, et al. Lentivirus pre-packed with Cas9 protein for safer gene editing. *Gene Ther*. 2016;23(7):627–633. https://doi.org/10.1038/gt.2016.27.
71. Aubrey BJ, Kelly GL, Kueh AJ, et al. An inducible lentiviral guide RNA platform enables the identification of tumor-essential genes and tumor-promoting mutations in vivo. *Cell Rep*. 2015;10(8):1422–1432. https://doi.org/10.1016/j.celrep.2015.02.002.
72. Persons DA. Lentiviral vector gene therapy: effective and safe? *Mol Ther*. 2010;18(5): 861–862. https://doi.org/10.1038/mt.2010.70.
73. Ramos-Kuri M, Rapti K, Mehel H, et al. Dominant negative Ras attenuates pathological ventricular remodeling in pressure overload cardiac hypertrophy. *Biochim Biophys Acta*. 2015;1853(11 pt A):2870–2884. https://doi.org/10.1016/j.bbamcr.2015.08.006.
74. Maggio I, Holkers M, Liu J, Janssen JM, Chen X, Gonçalves MAFV. Adenoviral vector delivery of RNA-guided CRISPR/Cas9 nuclease complexes induces targeted mutagenesis in a diverse array of human cells. *Sci Rep*. 2014;4:5105. https://doi.org/10.1038/srep05105.
75. Wang D, Mou H, Li S, et al. Adenovirus-mediated somatic genome editing of Pten by CRISPR/Cas9 in mouse liver in spite of Cas9-specific immune responses. *Hum Gene Ther*. 2015;26(7):432–442. https://doi.org/10.1089/hum.2015.087.
76. Wightman L, Kircheis R, Rössler V, et al. Different behavior of branched and linear polyethylenimine for gene delivery in vitro and in vivo. *J Gene Med*. 2001;3(4): 362–372. https://doi.org/10.1002/jgm.187.
77. Zuckermann M, Hovestadt V, Knobbe-Thomsen CB, et al. Somatic CRISPR/Cas9-mediated tumour suppressor disruption enables versatile brain tumour modelling. *Nat Commun*. 2015;6:7391. https://doi.org/10.1038/ncomms8391.
78. Zuris JA, Thompson DB, Shu Y, et al. Cationic lipid-mediated delivery of proteins enables efficient protein-based genome editing in vitro and in vivo. *Nat Biotechnol*. 2014;33(1):73–80. https://doi.org/10.1038/nbt.3081.
79. de Figueiredo IR, Freire JM, Flores L, Veiga AS, Castanho MA. Cell-penetrating peptides: a tool for effective delivery in gene-targeted therapies. *IUBMB Life*. 2014;66(3):182–194. https://doi.org/10.1002/iub.1257.
80. Ramakrishna S, Kwaku Dad A-B, Beloor J, Gopalappa R, Lee S-K, Kim H. Gene disruption by cell-penetrating peptide-mediated delivery of Cas9 protein and guide RNA. *Genome Res*. 2014;24(6):1020–1027. https://doi.org/10.1101/gr.171264.113.
81. Wang M, Zuris JA, Meng F, et al. Efficient delivery of genome-editing proteins using bioreducible lipid nanoparticles. *Proc Natl Acad Sci USA*. 2016;113(11):2868–2873. https://doi.org/10.1073/pnas.1520244113.
82. Yin H, Xue W, Chen S, et al. Genome editing with Cas9 in adult mice corrects a disease mutation and phenotype. *Nat Biotechnol*. 2014;32(6):551–553. https://doi.org/10.1038/nbt.2884.

83. Xue W, Chen S, Yin H, et al. CRISPR-mediated direct mutation of cancer genes in the mouse liver. *Nature*. 2014;514(7522):380–384. https://doi.org/10.1038/nature13589.
84. Bonamassa B, Hai L, Liu D. Hydrodynamic gene delivery and its applications in pharmaceutical research. *Pharm Res*. 2011;28(4):694–701. https://doi.org/10.1007/s11095-010-0338-9.
85. He N, Zeng X, Wang W, et al. Challenges and future expectations of reversed gene therapy. *J Nanosci Nanotechnol*. 2011;11(10):8634–8638.
86. Hruscha A, Schmid B. Generation of zebrafish models by CRISPR/Cas9 genome editing. *Methods Mol Biol*. 2015;1254:341–350. https://doi.org/10.1007/978-1-4939-2152-2_24.
87. Basu S, Aryan A, Overcash JM, et al. Silencing of end-joining repair for efficient site-specific gene insertion after TALEN/CRISPR mutagenesis in Aedes aegypti. *Proc Natl Acad Sci USA*. 2015;112(13):4038–4043. https://doi.org/10.1073/pnas.1502370112.
88. Khan KH. Gene transfer technologies and their applications: roles in human diseases. *Asian J Exp Biol Sci*. 2010;1(2):208–218.
89. Kim S, Kim D, Cho SW, Kim J, Kim J-S. Highly efficient RNA-guided genome editing in human cells via delivery of purified Cas9 ribonucleoproteins. *Genome Res*. 2014;24(6):1012–1019. https://doi.org/10.1101/gr.171322.113.
90. Kaneko T, Sakuma T, Yamamoto T, Mashimo T. Simple knockout by electroporation of engineered endonucleases into intact rat embryos. *Sci Rep*. 2014;4, 6382https://doi.org/10.1038/srep06382.
91. Straub C, Granger AJ, Saulnier JL, Sabatini BL. CRISPR/Cas9-mediated gene knockdown in post-mitotic neurons. *PLoS One*. 2014;9(8). e105584https://doi.org/10.1371/journal.pone.0105584.
92. Matano M, Date S, Shimokawa M, et al. Modeling colorectal cancer using CRISPR-Cas9–mediated engineering of human intestinal organoids. *Nat Med*. 2015;21(3):256–262. https://doi.org/10.1038/nm.3802.
93. Johnson JZ, Hockemeyer D. Human stem cell-based disease modeling: prospects and challenges. *Curr Opin Cell Biol*. 2015;37:84–90. https://doi.org/10.1016/j.ceb.2015.10.007.
94. González F, Zhu Z, Shi Z-D, et al. An iCRISPR platform for rapid, multiplexable, and inducible genome editing in human pluripotent stem cells. *Cell Stem Cell*. 2014;15(2):215–226. https://doi.org/10.1016/j.stem.2014.05.018.
95. Kotini AG, Chang C-J, Boussaad I, et al. Functional analysis of a chromosomal deletion associated with myelodysplastic syndromes using isogenic human induced pluripotent stem cells. *Nat Biotechnol*. 2015;33(6):646–655. https://doi.org/10.1038/nbt.3178.
96. Hendel A, Bak RO, Clark JT, et al. Chemically modified guide RNAs enhance CRISPR-Cas genome editing in human primary cells. *Nat Biotechnol*. 2015;33(9):985–989. https://doi.org/10.1038/nbt.3290.
97. Rieder F, Steininger C. Cytomegalovirus vaccine: phase II clinical trial results. *Clin Microbiol Infect*. 2014;20(suppl 5):95–102. https://doi.org/10.1111/1469-0691.12449.

CHAPTER SEVEN

Structural Simplicity and Mechanistic Complexity in the Hammerhead Ribozyme

Sara M. O'Rourke, William G. Scott[1]
The Center for the Molecular Biology of RNA, University of California at Santa Cruz, Santa Cruz, CA, United States
[1]Corresponding author: e-mail address: wgscott@ucsc.edu

Contents

1. Background and Structural Overview 178
2. Fast Minimal Hammerhead Ribozymes 181
3. Acid-Base Catalysis and the Hammerhead Ribozyme 182
4. Is the Hammerhead Ligation Reaction the Reverse of the Cleavage Reaction? 191
5. Do Cooperative Interactions in the Hammerhead Ribozyme Facilitate General Base Catalysis in the Cleavage Reaction? 194
6. Summary and Concluding Remarks 195
   6.1 The Structure of the Hammerhead Ribozyme May Be Much Simpler Than We Have Thought 196
   6.2 The Mechanism of the Hammerhead Ribozyme May Be Much More Complicated Than We Have Thought 196
   6.3 Concluding Remarks 200
References 201

Abstract

Natural or full-length hammerhead ribozymes are up to 1000-fold more active than their minimal counterparts that lack a complex tertiary interaction that pre-organizes and stabilizes the ribozyme active site, positioning RNA functional groups to facilitate acid-base catalysis. The recent discovery that a single tertiary contact (an AU Hoogsteen pair) between Stems I and II confers essentially all of the enhanced activity greatly simplifies our understanding of the structural requirements for hammerhead ribozyme activity. In contrast, the simplest mechanistic interpretations are challenged with the presentation of more complex alternatives. These alternatives are elucidated and critically analyzed in the context of several of the active hammerhead ribozyme structures now available.

Progress in Molecular Biology and Translational Science, Volume 159
ISSN 1877-1173
https://doi.org/10.1016/bs.pmbts.2018.07.006

1. BACKGROUND AND STRUCTURAL OVERVIEW

The hammerhead RNA (or hammerhead ribozyme) is a small, self-cleaving RNA motif found within various viroid and satellite RNA genomes, as well as RNA transcripts and retrotransposons. The name "hammerhead" originates from the conserved secondary structure, which apparently resembles the head of a ball-peen hammer (or those of the similarly maligned sharks in the family *Sphyrnidae*). The hammerhead, being a comparatively simple catalytic RNA sequence, is quite possibly the most intensively studied ribozyme, both from the point of view of mechanistic biochemical characterizations and structural investigations. In Chapter 1 of *Progress in Molecular Biology and Translational Science (vol. 120): RNA*, the structures of the minimal and full-length hammerhead ribozyme sequences, together with a discussion of their mechanisms and role in mammalian gene regulation, were presented.[1] In this chapter, we focus upon more recent developments, along with their implications for better understanding the mechanisms of cleavage and ligation, as well as ribozyme design. In this section, we will briefly summarize the most relevant points presented previously.

The so-called minimal hammerhead sequence is that which enables self-cleavage. Minimal hammerheads consist of a conserved core of 15 mostly invariant nucleotides that are flanked by three helical stems, as depicted in Fig. 1, and are characterized with a catalytic turnover rate constant (k_{obs}) of about 1/min.[2] About 15 years subsequent to characterization of the minimal hammerhead sequence, two research groups discovered that natural hammerhead RNAs always included an additional set of sequence elements, with very little apparent conservation, that were responsible for mediating a tertiary contact.[3,4] Several structures of hammerhead ribozymes with various tertiary contacts have now been elucidated.[5–8] In addition to revealing the details of a complex interaction between Stems I and II (Fig. 1), these structures reveal the catalytically relevant conformation of the conserved core residues in the hammerhead ribozyme's active site (Fig. 2). Presumably as a consequence of stabilizing the active-site conformation, full-length hammerhead ribozymes often possess a catalytic turnover rate constant (k_{obs}) greater than 100/min,[9] indicating that the presence of a natural tertiary contact in the ribozyme greatly enhances catalysis. Remarkably, the tertiary contact region evaded characterization for 15 years, due to the absence of any clear sequence conservation pattern.

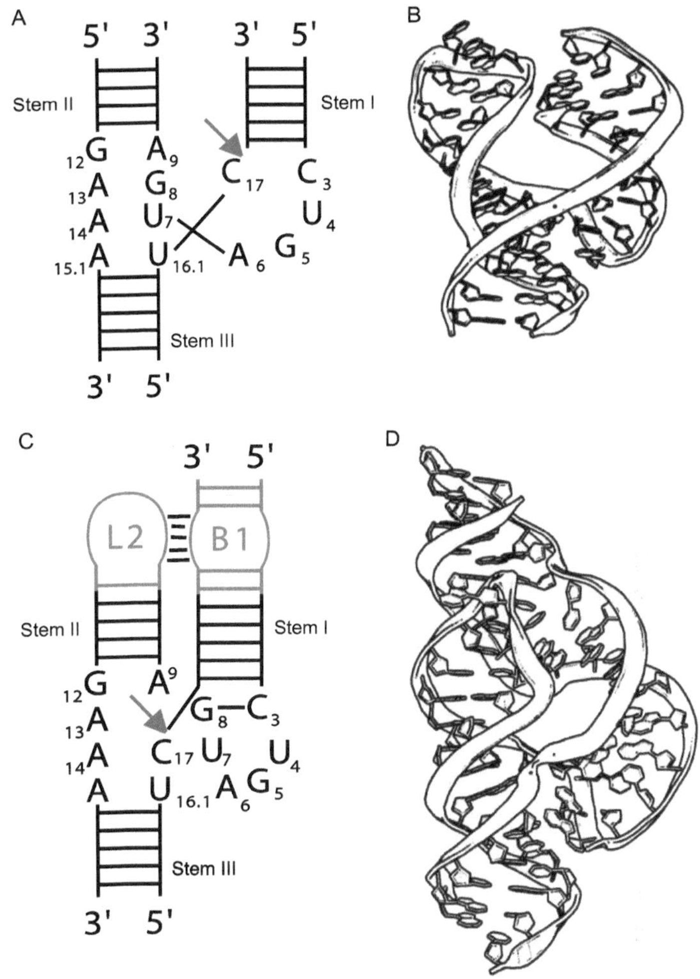

Fig. 1 The minimal and full-length hammerhead ribozymes. (A) A schematic representation of the secondary structure of the minimal hammerhead ribozyme. (B) The crystal structure of a corresponding minimal hammerhead ribozyme. The longer strand is the enzyme and the shorter strand is the substrate. (C) A schematic representation of the full-length hammerhead ribozyme emphasizing the presence of a tertiary contact between Stems I and II. (D) The crystal structure of a corresponding full-length hammerhead ribozyme. Again, the longer strand is the enzyme and the shorter strand is the substrate.

Both minimal and full-length hammerhead ribozyme sequences catalyze the same chemical reaction,[10] i.e., nucleophilic attack by a deprotonated 2′-O of the cleavage site residue (conventionally denoted position 17, and typically a C, although in addition to C17, U17- and A17-containing

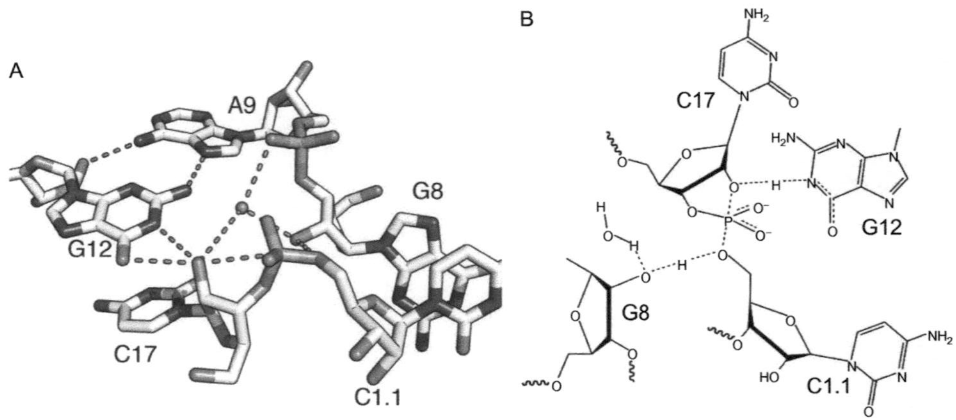

Fig. 2 The active site of the full-length hammerhead ribozyme and a mechanism for general acid-base catalysis. (A) Close-up of the crystal structure of the full-length hammerhead ribozyme showing G12 positioned for general base catalysis, the 2′-OH of G8 poised for acid catalysis, and the attacking nucleophile, the 2′-O of C17, positioned for an in-line attack upon the adjacent scissile phosphate. (B) A mechanistic diagram illustrating partial proton dissociation and transfer in a putative transition-state for general acid-base catalysis in the hammerhead ribozyme.

substrates are cleaved readily). G17 inhibits cleavage, presumably because it would have a strong propensity to pair with the invariant core residue C3, which is found in the full-length hammerhead structures to form a tertiary (but otherwise conventional) Watson-Crick pairing interaction with the invariant G8. (A potential G17-C3 pair would prevent the required G8-C3 pair from forming.) The hammerhead-catalyzed reaction is in essence the same as the first step of the two-step reaction catalyzed by RNase A, in which the deprotonated 2′-O of the cleavage site nucleophilically attacks the adjacent phosphorus, resulting in formation of cleavage products consisting of an RNA strand with a 2′,3′-cyclic phosphate terminus and another (the leaving group in the S_N2-like reaction) with a 5′-OH terminus. Based on the crystal structure[5] and several sets of biochemical analyses,[11] the invariant core residue G12 is thought to play a role analogous to His12 in RNase A, that is, it is believed to be the general base in the cleavage reaction (when in a deprotonated form). Similarly, the general acid functionality analogous to His119 appears to be provided by the 2′-OH of G8. The active-site structure and a suggested transition-state are depicted in Fig. 2.

2. FAST MINIMAL HAMMERHEAD RIBOZYMES

Minimal hammerhead ribozyme sequences are frequently used to design sequence-specific RNA cleavage reagents, primarily due to the lack of sequence restrictions of potential RNA targets, and also because no knowledge of the complexities of the three-dimensional structure of the hammerhead ribozyme is required to design a hammerhead to target the sequence of choice.[12] For example, any RNA sequence that possesses a nucleotide triplet of RUX (where R is G or A, and X is C, U or A) can be targeted by a hammerhead designed to base-pair with the targeted substrate sequence simply by designing complementary sequences to form Stems I and III flanking the cleavage site. Unfortunately, the minimal hammerhead is rather slow, especially under typical *in vivo* conditions (where the concentration of free Mg^{2+}, for example, is far below the 10 mM standard *in vitro* conditions that yield even modest turnover rates on the order of 1/min), which has frustrated design of potential therapeutic agents based upon minimal hammerhead sequences.

The discovery and characterization of the full-length, or natural, hammerhead sequence that includes the tertiary contact region offer the promise of the potential to design far more active hammerheads that will function efficiently under *in vivo* conditions. But because of the complex base-pairing and hydrogen bonding networks observed in the tertiary contacts of full-length hammerhead crystal structures, together with the apparent lack of any conserved sequence patterns (let alone the complexity of non-canonical base-pairing rules in tertiary interactions), designing full-length hammerheads to target arbitrary sequences of RNA has been considered far more difficult and daunting than the design of much less catalytically active minimal hammerhead sequences.

Recently, we discovered a very simple and highly effective work-around to the seemingly intractable problem of designing full-length hammerheads to cleave target RNAs.[13] We found that the *only* additional sequence restriction that needs to be placed upon the substrate RNA is to have a pyrimidine (preferably a U) at position 1.7, i.e., 7 nucleotides down-stream from the cleavage site. Hence, *any* RNA sequence of the form ...NNNNRUXNNNNNNYNN..., where N is any nucleotide, R is any purine, X is anything but G, and Y is any pyrimidine (preferably U), can be targeted with a fully active hammerhead sequence, provided that the 5′-end of the enzyme strand terminates at residue 2.6, and a GNRA tetraloop caps a Stem II sequence having exactly four base pairs. This permits the final A of the tetraloop to form a required Hoogsteen pair with U1.7 (Fig. 3). We experimentally determined that this Hoogsteen pair is both necessary and sufficient for conferring enhanced catalytic activity. All of the other tertiary contact pairing and hydrogen bonding interactions found in natural or full-length hammerhead sequences are not conserved, but are idiosyncratically required simply to prevent interactions forming that compete with this Hoogsteen pairing interaction. Design of synthetic hammerhead enzyme RNAs of a minimal sequence that terminates at residue 2.6 obviates the need for creating these additional contacts. As long as the pairing interaction can form between the A in the GNRA tetraloop and U1.7 in the substrate, such minimal hammerhead ribozyme sequences will possess the activity of full-length hammerheads, thus greatly simplifying the design of highly active RNA cleavage reagents.

3. ACID-BASE CATALYSIS AND THE HAMMERHEAD RIBOZYME

The first step of the RNase A cleavage reaction is the same as that catalyzed by the hammerhead ribozyme, i.e., nucleophilic attack of a 2′-O upon the adjacent phosphorus, resulting in formation of a 2′,3′-cyclic phosphate upon phosphodiester backbone cleavage and liberation of a

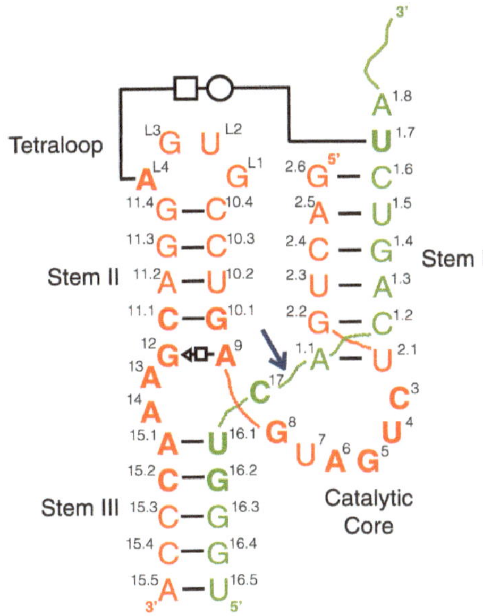

Fig. 3 A fast minimal hammerhead ribozyme, in which U1.7 of the substrate strand of Stem I can form a tertiary contact with the ultimate A of the GUGA tetraloop capping Stem II, via an AU Hoogsteen pairing interaction, indicated with a line marked with a circle and square. Conserved and invariant residues are indicated by bold-face. The phosphodiester cleavage site is indicated with an arrow. Stems I, II and III are labeled, as is the catalytic core region.

5′-OH RNA fragment downstream of the cleavage site. (The second step of the RNase A reaction, hydrolysis of the cyclic phosphate, does not have a hammerhead ribozyme counterpart.) As with RNase A, general acid-base catalysis is central to the enzyme mechanism. In the case of the RNase A cleavage reaction, the general base is His12 in the deprotonated form, and in the case of the hammerhead ribozyme, the general base appears to be G12 in the deprotonated form. The general acid in RNase A is a protonated His119, whereas the various full-length hammerhead ribozyme structures show the 2′-OH of G8 to be positioned for general acid catalysis in the cleavage reaction. It is therefore instructive to examine the pH-dependence of the RNA cleavage reactions catalyzed by RNase A and the hammerhead ribozyme to determine if these are consistent with the interpretation given.

For an enzyme that employs general acid-base catalysis in the rate-limiting step of the reaction, we can formulate the dependence of reaction rate upon pH using the Michaelis-Menten equation, recognizing that a general base in the active site of the enzyme will be inhibited by protonation, whereas a general acid in the active site will be inhibited by deprotonation (or activated by protonation). The most intuitive starting point is to assume competitive inhibition, so that the effect of protonation will alter the apparent K_M (K_M^{app} in Eq. 1). The effect of an enzyme inhibitor upon K_M is to add in a term proportional to the inhibitor concentration ([I] in Eq. 2) divided by the dissociation constant for the enzyme-inhibitor complex, and the effect of an activator is to add in a term proportional to the dissociation constant for the enzyme-activator complex divided by the concentration of the activator ([I'] in Eq. 2). In the case of general acid-base catalysis, both the activator and inhibitor are [H^+], the inhibitor dissociation constant *must be* the acid dissociation constant for the general base, denoted K_a^{GB}, because the general base is *inhibited* by increasing proton concentration, and the activator dissociation constant *must be* the acid dissociation constant for the general acid, K_a^{GA}, because the general acid is *activated* by increasing proton concentration, as shown in Eq. (3), which can be put into the more convenient form of Eq. (4) using the definitions of pH and pK_a:

$$v_o = \frac{V_{max}[S]}{K_M^{app} + [S]} \tag{1}$$

$$K_M^{app} = K_M\left(1 + \frac{[I]}{K_I} + \frac{K_{I'}}{I'}\right) \tag{2}$$

$$K_M^{app} = K_M\left(1 + \frac{[H^+]}{K_a^{GB}} + \frac{K_a^{GA}}{[H^+]}\right) \tag{3}$$

$$K_M^{app} = K_M\left(1 + \frac{10^{-pH}}{10^{-pK_a^{GB}}} + \frac{10^{-pK_a^{GA}}}{10^{-pH}}\right) \tag{4}$$

Returning to RNase A, the first step of the RNA cleavage reaction, the general base corresponds to the deprotonated form of His12, and the general acid corresponds to the protonated form of His119, allowing us to write the expression for K_{app} more explicitly (Eq. 5). Although both the general base and general acid are histidines, their actual pK_a values differ slightly due to differing microenvironments, with that of His12 being 5.8 and that of His119 being 6.2 (Eq. 6). The net effect of this in the context of the Michaelis-Menten equation for RNase A is a complex dependence of the initial rate upon pH (Eq. 7). Although we assumed competitive inhibition in this derivation, it is noteworthy that this assumption is not essential to the outcome. A similar dependence emerges for uncompetitive inhibition (Eq. 8), as well as for mixed inhibition:

$$K_M^{app} = K_M\left(1 + \frac{10^{-pH}}{10^{-pK_a^{His12}}} + \frac{10^{-pK_a^{His119}}}{10^{-pH}}\right) \quad (5)$$

$$K_M^{app} = K_M\left(1 + \frac{10^{-pH}}{10^{-5.8}} + \frac{10^{-6.2}}{10^{-pH}}\right) \quad (6)$$

$$v_o = \frac{V_{max}[S]}{K_M\left(1 + \frac{10^{-pH}}{10^{-5.8}} + \frac{10^{-6.2}}{10^{-pH}}\right) + [S]} \quad (7)$$

$$v_o = \frac{V_{max}[S]}{K_M + [S]\left(1 + \frac{10^{-pH}}{10^{-5.8}} + \frac{10^{-6.2}}{10^{-pH}}\right)} \quad (8)$$

To better focus upon the shape, rather than the magnitude, of the pH-dependence, we can set $[S] = K_M = V_{max} = 1$, and allow $[ES] = 1$. This enables us to plot pH vs. log k_{obs} in a manner essentially identical to that described previously by Bevilacqua.[14] The derivation used here, however, has an intuitive advantage in that it allows us to identify the explicit contribution of the general base, due to the inhibitory effect of increasing proton concentration (decreasing pH), and therefore also the contribution of the general acid (which is activated upon protonation). Fig. 4 shows the familiar pH-dependence of the observed rate constant (k_{obs}) for RNase A, with the maximum rate at pH 6.0. The log-linear increase in rate with increasing pH corresponds to deprotonation of His12, the general base, and the log-linear decrease in rate with increasing pH corresponds to deprotonation of His119,

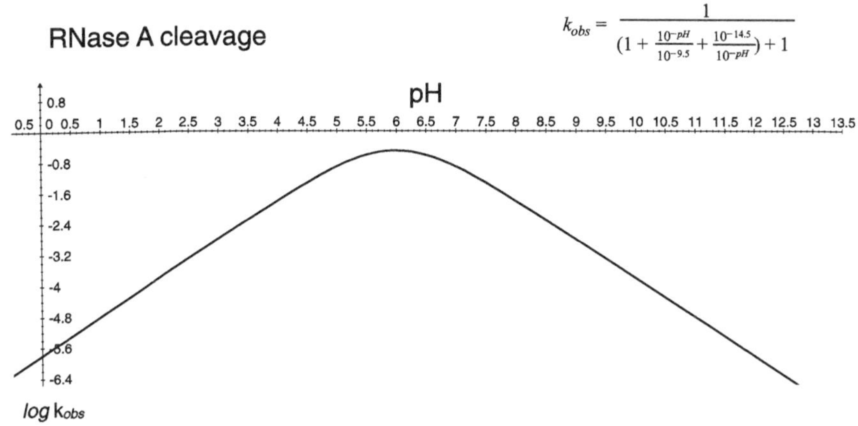

Fig. 4 A pH-dependent rate profile for RNase A.

the general acid. (The magnitude of the rate constant is arbitrary, and dependent upon parameters we have set to 1.0 for convenience, so this treatment, like that of Bevilacqua, focuses only upon relative rates.)

The hammerhead ribozyme also appears to employ general acid-base catalysis, wherein the deprotonated form of G12, hereafter referred to as $G12^-$, is thought to function in analogy with the deprotonated form of His12, based upon its position relative to the cleavage-site nucleotide in the various full-length hammerhead ribozyme structures. Unlike His, which has a physiologically important pK_a of 6.0, the relevant pK_a of guanosine is about 9.5. The predicted contribution to the relative rate is shown in Fig. 5, where the empirically observed pH-dependence is represented by black diamonds. Hence the hammerhead ribozyme rate profile is consistent with assignment of $G12^-$ as a general base. The apparent pK_a for the overall cleavage reaction reported in the literature varies between about 8.5 and 9.5 for the hammerhead ribozyme, depending upon the sequences used and reaction conditions, including the presence, identity and concentrations of divalent metal ions and other reaction additives, suggesting the possibility that the actual pK_a of G12 might be perturbed to a small extent, based upon metal binding to O^6 and N^7 of G12 (which has been observed in many crystal structures) and other microenvironment conditions.[15–18] Furthermore, substitution of G12 with non-natural purine nucleotides having altered pK_a values shifts the profile curve to the left (while decreasing the relative rate), consistent with the assignment of G12 to the role of general base[11,19] (Fig. 6).

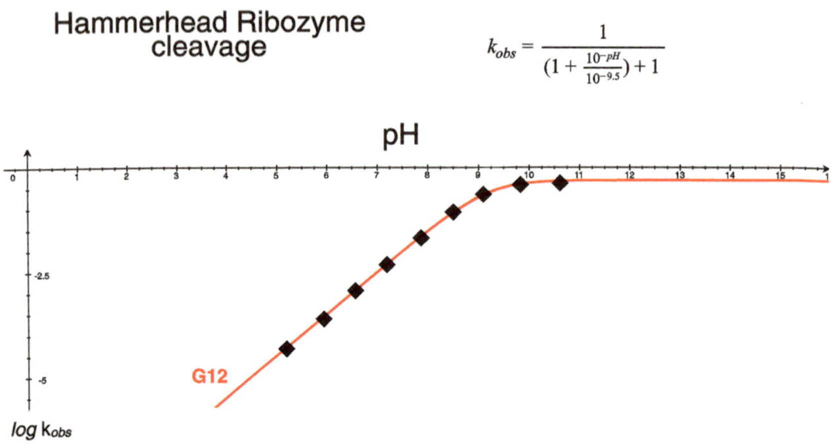

Fig. 5 A pH-dependent rate profile for G12-mediated general base-catalyzed hammerhead ribozyme cleavage, with representative empirically observed pH-dependent rate values represented as diamonds.

Based on crystal structures, the 2′-OH of G8 has been implicated as the general acid in the hammerhead cleavage reaction.[5] The unperturbed pK_a of a 2′-OH is about 14–15, which corresponds to a pH well above the range within which ribozyme kinetics experiments can be performed. Fig. 7 shows the theoretical contribution of the 2′-OH of G8 to the reaction rate profile, but all we can say empirically is that the resultant flat-topped curve (Fig. 8) is not inconsistent with the observed rate profile (superimposed black

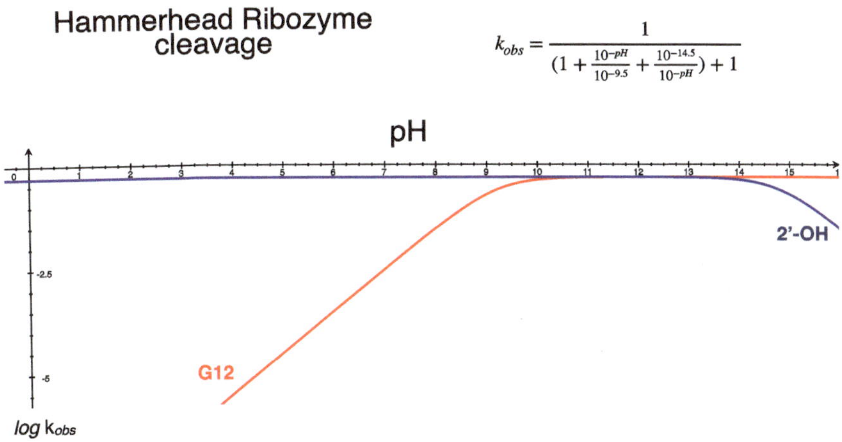

Fig. 6 A pH-dependent rate profile for G12 (general base) and 2′-OH (general acid) mediated hammerhead ribozyme cleavage.

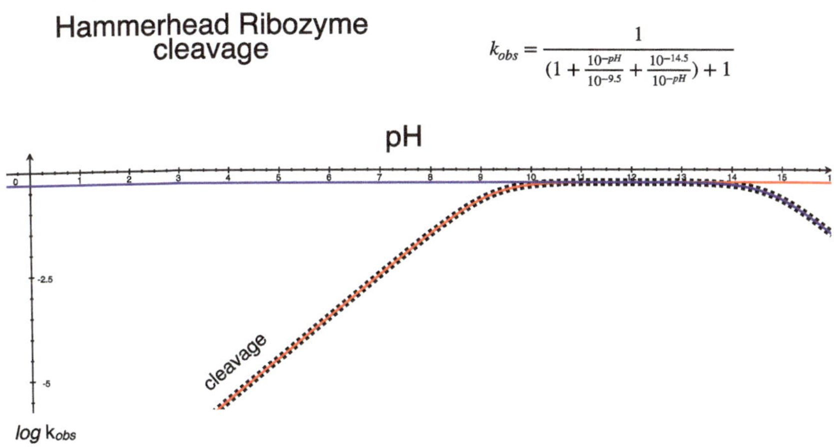

Fig. 7 A pH-dependent rate profile for G12 (general base) and 2′-OH (general acid) mediated hammerhead ribozyme cleavage, with the resultant theoretical pH-dependent rate plotted as a series of squares.

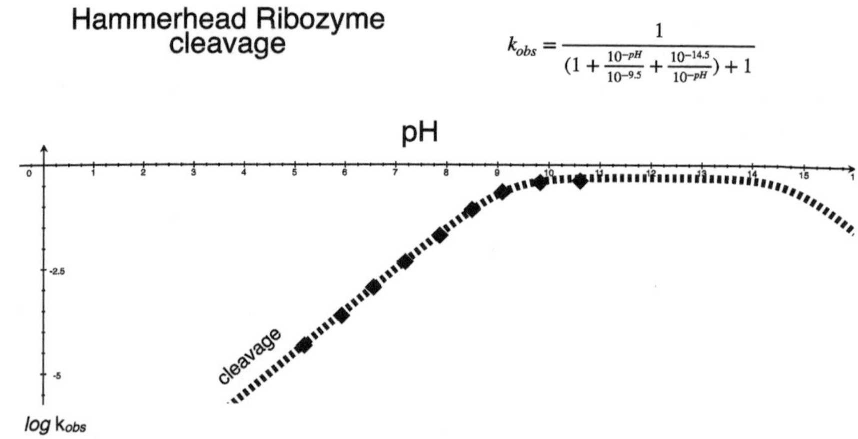

Fig. 8 The resultant from Fig. 7 (a series of squares) superimposed with the empirical observations from Fig. 5, showing that the general acid contribution to catalysis is not inconsistent with the observed rate dependence.

diamonds, as with Fig. 5). The plateau, which is experimentally observable for the hammerhead cleavage reaction under a variety of conditions,[20] results from the large gap between the pK_a for the general base (assumed to be 9.5 in these figures) and that of the general acid. Moreover, the plateau feature only emerges if the pK_a of the general base is much *lower* than that of the general acid. If the situation is reversed, wherein the pK_a of the general base is much *higher* than that of the general acid, not only is the relative rate of the reaction greatly diminished, but there is no plateau in the rate curve. Such a reversal would in fact occur for the hammerhead back reaction, because the roles of G12 and the 2′-OH of G8 would be reversed. The predicted pH-dependence of the ligation reaction, where G12 is now the general acid (with a pK_a of 9.5) and the 2′-OH of G8 is the general base, with a much higher pK_a of 14.5, is shown in Fig. 9. This prediction is also in accordance with the observation that the hammerhead ribozyme reaction, at least when measured with typical constructs and under typical reaction conditions, is one wherein cleavage dominates over the back reaction.[21] At physiological pH values, cleavage is in fact predicted to greatly dominate over the back reaction. The consequences of this are discussed further in Section 4.

The RNase A analysis referenced above was originally developed to help elucidate the mechanism of the hairpin ribozyme.[14] Like the hammerhead ribozyme, it catalyzes the same reaction as the first step of the RNase A cleavage reaction. The crystal structure, as well as other experiments, identifies $G8^-$ (in the deprotonated form) as the general base in the hairpin

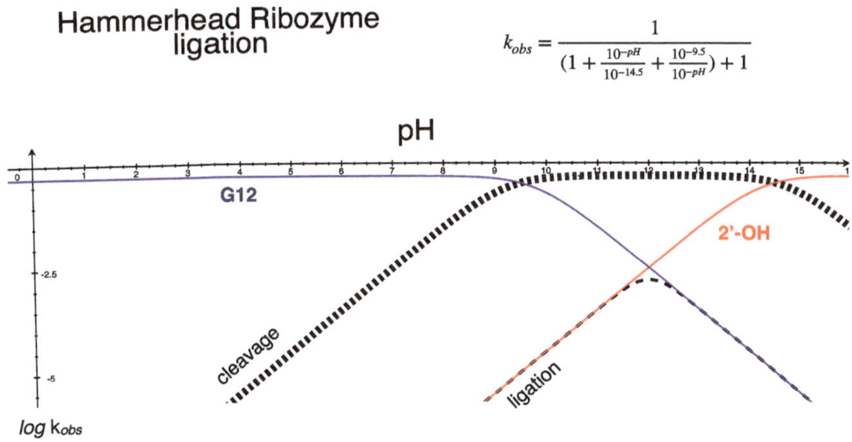

Fig. 9 Hammerhead ribozyme ligation, assuming that the mechanism is the reverse of the cleavage mechanism (i.e., that the 2′-OH is the general base, and G12 is the general acid). Ligation via the reverse-cleavage mechanism is thus predicted to be at least 1000-fold less efficient than cleavage at physiological pH.

ribozyme cleavage reaction, and A38$^+$ (in the protonated form) as the general acid. Unlike the hammerhead ribozyme, the hairpin ribozyme reaction has been well characterized as pH-independent throughout a large range of physiologically relevant pH values (i.e., between ~5 and ~8.5). The analysis employed for the RNase A reaction was also employed to examine the hairpin ribozyme cleavage reaction, reproducing the empirically observed pH-independent plateau region by combining titration curves for G8 and A38 (which have unperturbed pK_a values of 9.5 and 3.5, respectively).

The previous analysis of the hairpin ribozyme cleavage reaction has been repeated in Fig. 10. Consistent with the claim that the plateau feature only emerges if the pK_a of the general base is much *lower* than that of the general acid, no resultant plateau emerges from combining the titration curves for the general base, G8 and the general acid, A38. Instead, a curve resembling that of RNase A, but centered approximately at pH 6.5, is produced, which appears at first glance to be inconsistent with the observed dependence of k_{obs} upon pH. However, if one instead considers the ligation reaction, A38 becomes the general base, and G8 becomes the general acid. Fig. 11 shows that the expected plateau then emerges. Moreover, the ligation rate is predicted to dominate over the cleavage rate. This is gratifying, because the hairpin ribozyme is in fact observed to favor ligation. The result appears to be counter-intuitive unless one considers that $k_{obs} \neq k_{cleavage}$, but instead, $k_{obs} = k_{cleavage} + k_{ligation}$,[22] so that for a ribozyme in which ligase activity

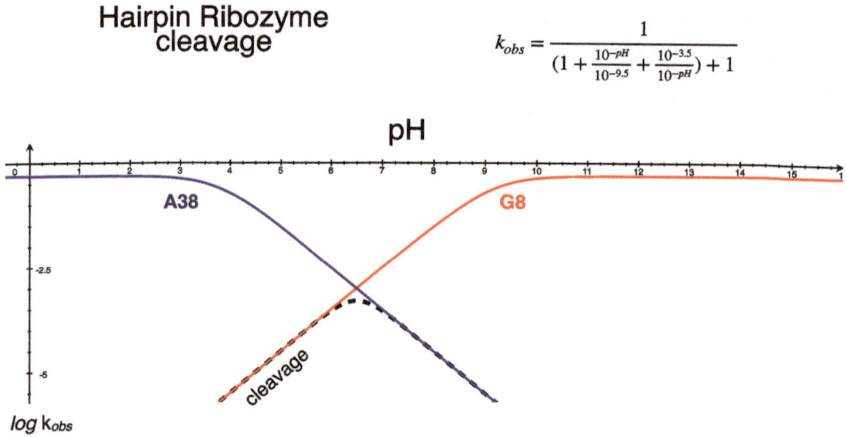

Fig. 10 The pH-dependence of the hairpin ribozyme cleavage reaction, with the hairpin ribozyme's G8 the general base, and A38 the general acid. Note the lack of a pH-independent plateau, in contrast with a previous analysis.

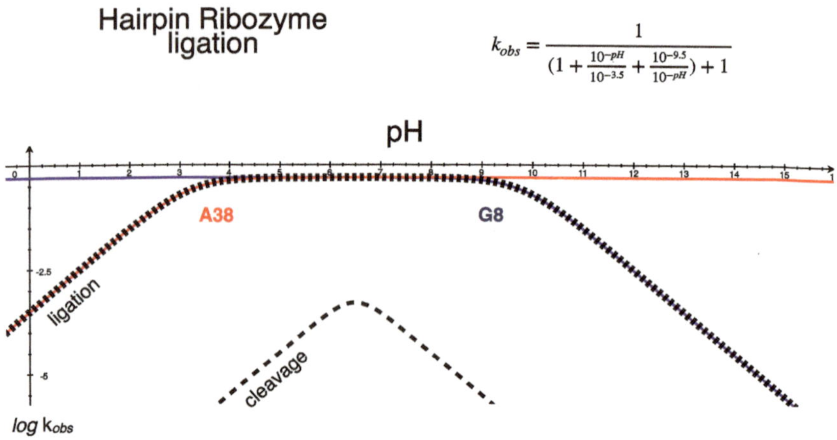

Fig. 11 The pH-dependence of the hairpin ribozyme ligation reaction, with the hairpin ribozyme's A38 the general base, and G8 the general acid. Note the pH-independent plateau is now a consequence of this analysis, and that the ligation reaction is favored over the cleavage reaction, in accordance with observation.

dominates, $k_{\text{obs}} \approx k_{\text{ligation}}$. In other words, standard single-turnover hairpin ribozyme kinetics analyses are not measuring the assumed cleavage reaction, but rather, they are (counter-intuitively) measuring the dominating ligation reaction. The approach to the derivation used here emphasizes the physical interpretation of the formalism, thus highlighting the consequences of the predictions, as well as the origins of the contributions from the general acid and general base, and, therefore, the direction of the reaction.

The hairpin ribozyme is believed to employ a pK_a shift in A38 (from 3.5 to 5.5). It is interesting to note the effect of this perturbation upon the rates of the cleavage and ligation reactions near neutral pH. In the case of the faster ligation reaction, the pK_a shift has no effect above pH 6, whereas it significantly enhances the slower cleavage rate between pH 6 and 8, and moves the pH of the maximum cleavage rate from about 6.5 to 7.5. By contrast, for the hammerhead ribozyme, the putative pK_a shift suggested for G12 (the cleavage reaction general base) would have the effect of further enhancing the rate of the cleavage reaction at physiological pH values, but at the expense of further attenuating the ligation reaction (albeit only at higher pH values).

4. IS THE HAMMERHEAD LIGATION REACTION THE REVERSE OF THE CLEAVAGE REACTION?

The revised analysis for the hairpin ribozyme depicted in Fig. 12 clearly demonstrates how A38 and G8 can effectively collaborate to efficiently catalyze both the cleavage and ligation reactions required in the context of satellite virus RNA rolling circle replication, and how a suggested pK_a perturbation of A38 further enhances this potential by narrowing the difference between the ligation and cleavage rates. In contrast, the hammerhead cleavage reaction is very efficient, but the back reaction is drastically

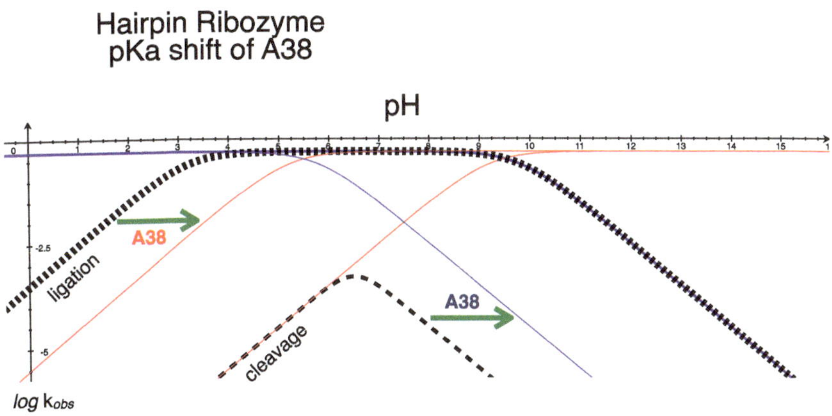

Fig. 12 The effects of a putative pK_a shift of A38 in the hairpin ribozyme cleavage and ligation reactions. Note that at physiological pH values, the shift only effects the cleavage reaction.

less so, especially at physiological pH (Fig. 9). Because ligation activity is also required for the reproductive cycle of satellite RNAs that have embedded hammerhead sequences, our analysis suggests that the hammerhead might be far less favored by natural selection than the hairpin ribozyme, an efficient ligase, whose nuclease activity is also appreciable. Nevertheless, hammerhead sequences are much more common than hairpin sequences,[23,24] suggesting that there might be more to this story, i.e., that the simple, straightforward explanation that the ligation reaction is simply the reverse mechanism of the cleavage reaction clearly lacks explanatory power. In addition, Class I hammerheads, such as the one discovered in *Schistosoma mansoni* circular RNA elements, and the closely related Clec2-associated hammerheads found in the 3′-UTRs of mammalian mRNAs,[25] exhibit substantial ligase activity under favorable conditions,[26] seemingly at odds with the explanation that the ligation reaction is simply the reverse mechanism of the cleavage reaction.

Consistent with the prediction of two separate mechanisms for cleavage and ligation, respectively, is the case of *Avsunviroidae* viroids, in which the ligation reaction appears to be catalyzed by the chloroplastic isoform of tRNA ligase, a protein enzyme that uses 5′-OH and 2′,3′-cyclic phosphate termini as a substrate.[27] It is unclear whether other hammerhead ligation reactions are also catalyzed by tRNA ligases. In addition, if the hammerhead ribozyme evolutionarily predated tRNA ligases (which are involved in the last step of spliceosome-independent intron excision in several tRNAs),[28] it would imply that hammerhead (or hammerhead-like) sequences would have had to possess the capacity to catalyze ligation in the absence of tRNA ligase.

One possibility is that the ligation reaction of the hammerhead ribozyme is mechanistically distinct from the cleavage reaction. In other words, since the 2′-OH of G8 has such a limited chemical potential as a general base, some other functionality within the hammerhead RNA (instead of external functionality provided by tRNA ligase) might instead participate in the ligation reaction, possibly requiring a conformational change in the hammerhead ribozyme structure in order to do so. The same studies that successfully identified G12 as a participant in general acid/base catalysis prior to the availability of the full-length hammerhead structure also implicated G8.[11] (Although the fact that the putative general acid in the cleavage reaction is the 2′-OH of G8, this may only be coincidental, as this previous study implicated the purine nucleotide base itself.) Comparing the minimal hammerhead structure to the full-length hammerhead structure is facilitated by adiabatic morphing between the two sets of crystallographic coordinates, which also demonstrates that the minimal hammerhead conformation can be continuously deformed until it arrives at the observed conformation of the full-length hammerhead.[29] The most extreme change in the active site is the dramatic movement of G8, which in the minimal hammerhead structure is stacked under A9 but unpaired and that forms a tertiary Watson-Crick base-pairing interaction with C3 in the full-length hammerhead structure. The adiabatic morphing procedure interpolates an energetically plausible but hypothetical trajectory between the two crystallographic structural endpoints, generating 50 intermediates along the path of the conformational rearrangement. Intermediate #40 is one in which G8 is unstacked with respect to A9 but has not yet formed hydrogen bonds with C3. Instead, the Watson-Crick face of G8 points directly at the scissile phosphate. Superposition with the cleaved structure (2quw) suggests that the N1 of G8 in intermediate #40 may in fact be well positioned to facilitate general base catalysis in the ligation reaction, as shown in Fig. 13, implying that the observed plasticity of the hammerhead ribozyme, emulated in the adiabatic morphing procedure, could facilitate separate mechanisms for the cleavage and ligation reactions.

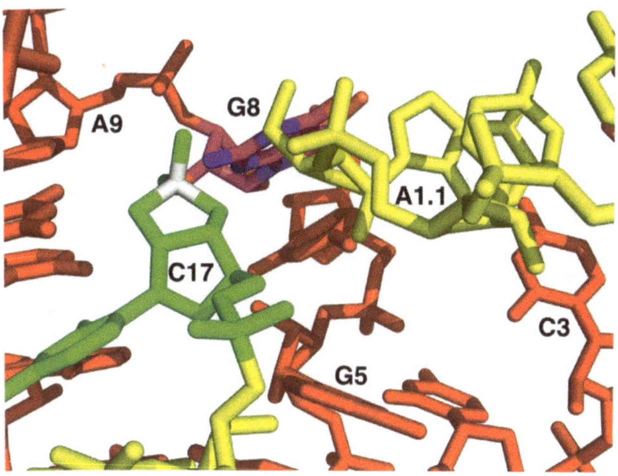

Fig. 13 Intermediate step #40 in the adiabatic morphing simulation of the conformational transition between minimal and full-length hammerhead ribozyme structures positions the N^1 of G8 proximal to the 5′-OH of what will be the attacking nucleophile in the ligation reaction of a previously cleaved hammerhead ribozyme, in which the 5′-OH-terminated and 2′,3′-terminated ends of two RNA fragments will be ligated. Based upon this structural model, as well as biochemical experiments that implicate the nucleobase of G8 as mechanistically important, we propose that G8 becomes unpaired from C3 to serve as a general base in a mechanistically distinct ligation reaction.

Although it may be objected that this proposal appears to violate the principle of microscopic reversibility, it is worth remembering that this principle applies only to concerted, one-step reactions, whereas the hammerhead reaction, by necessity, will involve more than one step, as the initial protonation states of the enzyme must be restored subsequent to the cleavage reaction, and the division of the hammerhead RNA into distinct enzyme and substrate strands is an artificial experimental construct; the true biological reaction involves a single cleavage event followed by a single ligation event. In addition, a tRNA ligase-catalyzed reverse reaction is in a sense the most extreme example of a distinct mechanism for the reverse reaction. The proposal that at least some hammerhead ribozymes might possess a mechanistically distinct reverse reaction should be testable in that it predicts altered rate-pH profiles for purines substituted for G8 in constructs having a compensatory substitution at C3 that maintains Watson-Crick base-pairing (and thus the shape of the rate-pH profile for the cleavage reaction).

5. DO COOPERATIVE INTERACTIONS IN THE HAMMERHEAD RIBOZYME FACILITATE GENERAL BASE CATALYSIS IN THE CLEAVAGE REACTION?

Two recent publications argue that even the comparatively well-understood mechanism for how the general base facilitates catalysis in the cleavage reaction of the hammerhead ribozyme may in fact be far more complex than previously appreciated.[30,31] It is proposed that the cleavage-site nucleotide, C17, becomes protonated, and therefore positively charged, at physiological pH, and that in the transition-state, cation-π interactions between $G12^-$ (the deprotonated general base), $C17H^+$ (the protonated cleavage-site nucleotide), and U16.1 (the invariant nucleotide 5′ to C17), cooperatively stabilize the catalytically relevant form of the general base.

Chelation of a metal ion via O6 and N7 on the Hoogsteen face of G12 is thought to lower the apparent pK_a of G12, making it a more potent general base.[32,33] This interaction, which has been frequently observed[7,8,34–42] with both monovalent and divalent cations (e.g., Na^+, Mg^{2+}, Mn^{2+}), may help to stabilize the deprotonated state ($G12^-$) required for G12 to play the role of the general base in the cleavage reaction (in analogy to the requirement for a deprotonated His12 in RNase A). Stabilization of the deprotonated state ($G12^-$) would thus manifest itself as an apparent pK_a shift that is dependent upon the presence of a bound metal ion on the face of G12 and could help to explain the observation that metal ions enhance catalysis while not being strictly required for the catalytic mechanism.[43]

Under physiological conditions in which very little (~1 mM) free Mg^{2+} is present, C17H$^+$, the protonated form of the cleavage-site nucleotide, is proposed to provide the positive charge in the absence of a metal ion chelated to G12. C17 in this context is proposed to experience a pronounced pK_a shift toward neutrality that enables it to facilitate catalysis by stabilizing the deprotonated state (G12$^-$) of the general base in a cooperative manner.[30] Experiments employing a kinetically well-characterized minimal hammerhead ribozyme reveal pH-dependent rate profiles that show not only the anticipated pK_a shift in the presence of increasing divalent metal ion concentrations, but also reveal "wavy" features in the rate profile curve at higher (e.g., ~50 mM) concentrations of Mg^{2+} due to data points that deviate from a simple G12 titration curve. These features are interpreted as evidence for a cooperative interaction that takes place between G12$^-$ and C17H$^+$, and the observed phenomenon is subsequently generalized to an elaborate, multichannel three-dimensional kinetics scheme involving "dark pK_a shifts and wavy rate-pH profiles"[31] to describe general acid-base catalysis.

6. SUMMARY AND CONCLUDING REMARKS

The mechanism describing hammerhead ribozyme activity should be able to account for (1) general base catalysis of the cleavage reaction, (2) general acid catalysis of the cleavage reaction, (3) general base catalysis of the ligation reaction, (4) general acid catalysis of the ligation reaction, (5) stabilization of the conformations of the enzyme and substrate in a pre-arranged active site, and (6) stabilization of the excess negative charge that accumulates in the pentacoordinated oxyphosphorane transition-state. The current state of our understanding can be summarized in these terms. (1) G12$^-$ is almost certainly the general base in the cleavage reaction, (2) the 2′-OH of G8 may be the general acid in the cleavage reaction, (3) the general base in the ligation reaction would therefore be the 2′-OH of G8, which is problematic, as noted above, (4) the general acid in the ligation reaction would therefore be G12, (5) the full-length hammerhead ribozyme crystal structures reveal the pre-organized active site of the hammerhead ribozyme and how it is stabilized by the tertiary contact between Stems I and II, and the possible involvement of a metal ion in transition-state stabilization (6) is still a matter of active debate, with a suggestion (for example) that the A9 and scissile phosphates both coordinate a single metal ion in the transition-state. Only the first five aspects are examined here.

6.1 The Structure of the Hammerhead Ribozyme May Be Much Simpler Than We Have Thought

The essential structural principles of the hammerhead ribozyme may be far less complex than the various full-length hammerhead RNA crystal structures first suggested. In particular, the only required feature of the complicated network of hydrogen bonds and non-canonical tertiary base-pairing interactions between Stems I and II is the single AU Hoogsteen pairing between the A in the GNRA tetraloop of Stem II, and U1.7 in the substrate strand of Stem I. A minimal hammerhead RNA that permits this required interaction to form, by virtue of its design, possesses all of the catalytic enhancement of the full-length natural hammerhead sequences without additional complexities introduced by the unconserved interactions comprising the rest of the tertiary contact.

6.2 The Mechanism of the Hammerhead Ribozyme May Be Much More Complicated Than We Have Thought

The cleavage reaction catalyzed by the hammerhead ribozyme is rather simple. Abstraction of the 2′-proton from the 2′-OH of the cleavage-site nucleotide generates a nucleophile that attacks the adjacent phosphorus, breaking the 3′–5′ phosphodiester linkage while forming a 2′,3′-cyclic phosphodiester, and generating a 5′-OH-terminated RNA strand as the leaving group in an S_N2 reaction as the emerging 5′-O^- is protonated. In biochemistry, it does not get much simpler. The reaction is thought to be catalyzed by general acid-base catalysis, in a manner analogous to that of the first step of the RNase A reaction. $G12^-$ is thought to be the general base (in analogy to deprotonated His12 in RNase A). A 2′-OH in the enzyme strand (belonging to G8) is positioned within hydrogen bonding distance of the leaving group 5′-O in the various crystal structures, leading to the suggestion that it may supply a proton (perhaps in a water-mediated proton relay), thus serving as a general acid. In that sense, the reaction should be no more complicated than that of RNase A, and in fact simpler, since the hammerhead only catalyzes the first step of the RNase A reaction (unlike RNase A, it does not hydrolyze the 2′,3′-cyclic phosphodiester).

The principle of parsimony (attributed originally to William of Occam) is a heuristic edict suggesting that the simplest hypothesis explaining a phenomenon should be favored (and is generally the most potent argument against conspiracy theories). In science, it places the burden of proof upon the proponents of more complex explanations, and this burden should include providing evidence that a simpler explanation is insufficient, and that the more elaborate explanation or theory has greater explanatory power. Therefore, any proposal for a more complicated enzyme mechanism must begin with a compelling argument for why simpler alternatives lack explanatory power.

6.2.1 The Ligation Reaction Mechanism Might Not Be the Reverse of the Cleavage Mechanism

General base catalysis by $G12^-$, by itself, is sufficient to explain the observed pH-dependent rate profile of the hammerhead ribozyme cleavage reaction. Although the 2′-OH of G8 implicated as the general acid is not inconsistent with the observed pH-dependence, it is not strictly speaking required to explain the observed data. Specific-acid catalysis (i.e., protonation of the leaving group oxygen via a hydronium ion), for example, is equally consistent with the observations. All that can really be said with confidence is that the acid catalyst in the cleavage reaction has a very high pK_a.

The Principle of Microscopic Reversibility states that, for a concerted (or single-step) chemical reaction, the transition-state for the forward and reverse reactions is the same. This means that the mechanism for a concerted forward reaction should be the reverse of the back reaction. For this reason, the parsimonious expectation is that the hammerhead ligation reaction is simply the reverse of the cleavage reaction.

The replicative cycle of the satellite virus RNAs that contain hammerhead ribozymes requires both efficient cleavage activity and efficient ligation activity. However, as shown in Section 3, ligation via the reverse of the cleavage mechanism will be extremely inefficient, and thus lacks the power to explain how the satellite virus RNA can replicate. This in turn compels us to propose that hammerhead RNA ligation may take place via a mechanism that is distinct from the cleavage mechanism. Indeed, a separate (protein) enzyme, tRNA ligase, catalyzes the reverse reaction in at least a subset of hammerhead-containing viroid RNAs.[27] Can a conformationally altered form of the hammerhead ribozyme itself do something analogous to this? Based on an intermediate structure interpolated via an adiabatic morphing of the minimal hammerhead RNA crystal structure conformation to that of the full-length hammerhead crystal structure, as well as its implication in biochemical experiments, we suggest that the nucleobase of the invariant nucleotide G8 (in its N^1-deprotonated form, $G8^-$) may be the base in the ligation reaction, and G12 (in its protonated form) continues to play its expected roll as the general acid in the ligation reaction (i.e., in exactly the same way as it would be expected to in the reverse-cleavage mechanism). We therefore propose an experimentally testable ligation mechanism that differs from the reverse of the cleavage mechanism only to the degree required by empirical observation. We note that although this proposal contradicts the parsimonious spirit of Microscopic Reversibility, it does not in fact contradict the actual letter of this Principle in that acid/base catalysis can never be a true concerted reaction. (Regeneration of the correct protonation states of the acid and base is a required second step, as can be seen from examination of the mechanism of RNase A, in which the hydrolysis of the 2′,3′-cyclic phosphate in the second step has the effect of doing exactly this.)

6.2.2 General Base Catalysis in the Cleavage Reaction Mechanism Might Be More Complex

To explain the functional form of the pH-dependent rate profile observed at higher Mg^{2+} concentrations, it is proposed that the cleavage-site nucleotide, C17, becomes protonated, and therefore positively charged, at physiological pH, and that in the transition-state, cation-π interactions between $G12^-$ (the deprotonated general base), $C17H^+$ (the protonated cleavage-site nucleotide), and U16.1 (the invariant nucleotide 5′ to C17) cooperatively stabilize the catalytically relevant form of the general base.[30]

Although this is a quite intriguing mechanistic proposal that possesses the merit of recognizing that the hammerhead ribozyme structure does more than provide a passive scaffold that positions metal ions for RNA catalysis, the invoked "dark pK_a shifts" and "wavy rate-pH profiles"[31] introduce a large degree of complexity. We need to ask whether this complexity is accompanied by a corresponding advance in explanatory power of how the hammerhead ribozyme works. In terms of the six aspects of the hammerhead reaction mechanism, this proposal most clearly addresses point 1 (general base catalysis in the cleavage reaction) in that it explains "wavy" rate profiles but may also be relevant to point 4 (general acid catalysis in the ligation reaction) in that a protonated $C17^+$ that favors $G12^-$ would also disfavor the protonated G12. This mechanistic proposal thus appears to have the unintended consequence of possibly gaining some more explanatory power in terms of the cleavage reaction at the expense of losing even more explanatory power in terms of the ligation reaction.

It is also possible that the deviation from an ideally shaped general base titration curve observed in the minimal hammerhead is an artifact due to the use of a minimal hammerhead ribozyme for these experiments. High concentrations of metal ions may simply facilitate close approach of Stems I and II to partially mimic the effects of the tertiary contact that cannot form in this minimal construct. If the deviation is found to be statistically significant and reproducible in full-length hammerhead ribozymes that naturally form a tertiary contact, the need to invoke a complex cooperative interaction would be far more compelling.

Fig. 5 illustrates that general base catalysis (alone) by G12, with an unshifted pK_a of 9.5, has a high degree of explanatory power in that its titration curve reproduces the pH-dependence of the hammerhead ribozyme cleavage rate within the experimentally relevant window between pH 5 and 10.5. As noted in Section 3, the identity and presence of a metal ion bound to the Hoogsteen face of G12 can slightly alter the apparent pK_a of the reaction profile (i.e., it can influence the inflection point of the curve without altering its shape), and that substitution of G12 with purines having lower pK_a values is also consistent with this interpretation. It is also apparent from Fig. 5 that a small ($\approx$1 pK_a unit) shift will have a comparatively minor relative cleavage rate enhancement, compared to the rather large rate enhancement of the hairpin ribozyme cleavage rate (Fig. 12) due to a proposed pK_a shift in A38. Invoking even a (noncooperative) pK_a shift in G12 offers little additional explanatory power. In addition, the experiments in which G12 is substituted with purines having lower pK_a values uniformlyresult in less active hammerhead ribozymes,[11] presumably because even though a greater fraction of the G12-substituted purine is in the deprotonated state at a given pH, it is a correspondingly less potent general base. Although a cation proximal to G12 may analogously change its effective pK_a, it does not entail that G12 will be a more potent general base.

The magnitude of the proposed pK_a shift in C17 is claimed to be significantly larger than the magnitude of the G12 pK_a shift thought to enhance hammerhead ribozyme catalysis.[30] Therefore, if protonation of C17 is required in the absence of a G12$^-$ stabilizing divalent cation, one has substituted the (relatively simple) requirement for understanding how a small pK_a shift takes place in G12 with the (more complex) requirement for understanding how a much larger pK_a shift takes place in C17. Since U17 can be substituted for C17, we now have the additional complexity of explaining how a nucleotide base that never possesses a positive charge is tolerated in natural hammerhead sequences.

In contrast, the natural *Schistosomal* hammerhead RNA sequence gives rise to a tertiary contact structure in which two cytosines form a base-pair,[38] requiring an extra proton to be present at neutral pH (i.e., the crystal structure is consistent with a large pK_a shift of the tertiary contact cytosine). Despite the requirement for the presence of the tertiary contact for the observed (cooperative) 10^3 rate enhancement of this hammerhead sequence,[9,42] there is no apparent signature in the rate-pH profile. Likewise, in RNase A, Lysine 41 (with a pK_a of approximately 10) does not influence the shape of the symmetric RNase A rate-pH profile, despite its close physical proximity to His12 and His119. These counter-examples should serve as positive controls when characterizing the empirical consequences of "dark" pK_a shifts, and evaluating the veracity of the claim.

6.3 Concluding Remarks

The hammerhead ribozyme, like all enzymes, catalyzes both a forward (cleavage) and reverse (ligation) reaction. Although the full-length hammerhead ribozyme is a much more efficient catalyst compared to the minimal hammerhead, at first glance this efficiency appears to come at the price of structural complexity. However, a more careful analysis reveals an underlying simplicity that had been masked by the presence of an elaborate tertiary structure, almost all of which is not strictly required for catalytic enhancement. This insight offers not only a great simplification in terms of understanding how the ribozyme works from a structural perspective, but also greatly simplifies the design of highly active hammerhead ribozymes as potential cleavage reagents that target RNA sequences of choice.

Conversely, the mechanism of the hammerhead ribozyme reaction may be far more complicated than originally assumed. The catalytic strategy employed by the hammerhead to accelerate the cleavage reaction appears to defy the need for efficient ligation. In a subset of natural hammerhead RNA sequences, a protein enzyme (tRNA ligase) appears to fill this need, but another subset of hammerheads are observed to catalyze their own ligation reaction efficiently, which raises the question of whether a different mechanistic strategy is employed for the ligation reaction. Based upon our previous structural analyses, we have suggested one such possible mechanism and have proposed how it might be tested experimentally.

REFERENCES

1. Scott WG, Horan LH, Martick M. The hammerhead ribozyme: structure, catalysis, and gene regulation. *Prog Mol Biol Transl Sci*. 2013;120:1–23.
2. Blount KF, Uhlenbeck OC. The structure-function dilemma of the hammerhead ribozyme. *Annu Rev Biophys Biomol Struct*. 2005;34:415–440.
3. De la Peña M, Gago S, Flores R. Peripheral regions of natural hammerhead ribozymes greatly increase their self-cleavage activity. *EMBO J*. 2003;22:5561–5570.
4. Khvorova A, Lescoute A, Westhof E, Jayasena SD. Sequence elements outside the hammerhead ribozyme catalytic core enable intracellular activity. *Nat Struct Biol*. 2003;10:708–712.
5. Martick M, Scott WG. Tertiary contacts distant from the active site prime a ribozyme for catalysis. *Cell*. 2006;126:309–320.
6. Chi Y-I, Martick M, Lares M, Kim R, Scott WG, Kim S-H. Capturing hammerhead ribozyme structures in action by modulating general base catalysis. *PLoS Biol*. 2008;6, e234.
7. Mir A, Golden BL. Two active site divalent ions in the crystal structure of the hammerhead ribozyme bound to a transition state analogue. *Biochemistry*. 2016;55:633–636.
8. Mir A, Chen J, Robinson K, et al. Two divalent metal ions and conformational changes play roles in the hammerhead ribozyme cleavage reaction. *Biochemistry*. 2015;54: 6369–6381.
9. Canny MD, Jucker FM, Kellogg E, Khvorova A, Jayasena SD, Pardi A. Fast cleavage kinetics of a natural hammerhead ribozyme. *J Am Chem Soc*. 2004;126:10848–10849.
10. Nelson JA, Uhlenbeck OC. Minimal and extended hammerheads utilize a similar dynamic reaction mechanism for catalysis. *RNA*. 2008;14:43–54.
11. Han J, Burke JM. Model for general acid-base catalysis by the hammerhead ribozyme: pH-activity relationships of G8 and G12 variants at the putative active site. *Biochemistry*. 2005;44:7864–7870.
12. Eckstein F, Aurup H, Benseler F, et al. Design of hammerhead ribozymes for the inhibition of gene expression ex vivo by exogenous application. *Nucleic Acids Symp Ser*. 1993;29:115.
13. O'Rourke SM, Estell W, Scott WG. Minimal hammerhead ribozymes with uncompromised catalytic activity. *J Mol Biol*. 2015;427:2340–2347.
14. Bevilacqua PC. Mechanistic considerations for general acid-base catalysis by RNA: revisiting the mechanism of the hairpin ribozyme. *Biochemistry*. 2003;42:2259–2265.
15. Dahm SC, Derrick WB, Uhlenbeck OC. Evidence for the role of solvated metal hydroxide in the hammerhead cleavage mechanism. *Biochemistry*. 1993;32:13040–13045.
16. Nelson JA, Uhlenbeck OC. Minimal and extended hammerheads utilize a similar dynamic reaction mechanism for catalysis. *RNA*. 2008;14:43–54.
17. Nelson JA, Uhlenbeck OC. Hammerhead redux: does the new structure fit the old biochemical data? *RNA*. 2008;14:605–615.
18. Shepotinovskaya IV, Uhlenbeck OC. Catalytic diversity of extended hammerhead ribozymes. *Biochemistry*. 2008;47:7034–7042.
19. Schultz EP, Vasquez EE, Scott WG. Structural and catalytic effects of an invariant purine substitution in the hammerhead ribozyme: implications for the mechanism of acid-base catalysis. *Acta Crystallogr D Biol Crystallogr*. 2014;70:2256–2263.
20. Hertel KJ, Herschlag D, Uhlenbeck OC. A kinetic and thermodynamic framework for the hammerhead ribozyme reaction. *Biochemistry*. 1994;33:3374–3385.
21. Hertel KJ, Uhlenbeck OC. The internal equilibrium of the hammerhead ribozyme reaction. *Biochemistry*. 1995;34:1744–1749.
22. Stage-Zimmermann TK, Uhlenbeck OC. Hammerhead ribozyme kinetics. *RNA*. 1998;4:875–889.

23. de la Peña M, Cervera A. Circular RNAs with hammerhead ribozymes encoded in eukaryotic genomes: the enemy at home. *RNA Biol.* 2017;14:985–991.
24. de la Peña M, García-Robles I, Cervera A. The hammerhead ribozyme: a long history for a short RNA. *Molecules.* 2017;22:78.
25. Martick M, Horan LH, Noller HF, Scott WG. A discontinuous hammerhead ribozyme embedded in a mammalian messenger RNA. *Nature.* 2008;454:899–902.
26. Canny MD, Jucker FM, Pardi A. Efficient ligation of the Schistosoma hammerhead ribozyme. *Biochemistry.* 2007;46:3826–3834.
27. Nohales MÁ, Molina-Serrano D, Flores R, Daròs JA. Involvement of the chloroplastic isoform of tRNA ligase in the replication of viroids belonging to the family Avsunviroidae. *J Virol.* 2012;86:8269–8276.
28. Abelson J, Trotta CR, Li H. tRNA splicing. *J Biol Chem.* 1998;273:12685–12688.
29. Scott WG. Morphing the minimal and full-length hammerhead ribozymes: implications for the cleavage mechanism. *Biol Chem.* 2007;388:727–735.
30. Frankel EA, Strulson CA, Keating CD, Bevilacqua PC. Cooperative interactions in the hammerhead ribozyme drive pK_a shifting of G12 and its stacked base C17. *Biochemistry.* 2017;56:2537–2548.
31. Frankel EA, Bevilacqua PC. Complexity in pH-dependent ribozyme kinetics: dark pK_a shifts and wavy rate-pH profiles. *Biochemistry.* 2018;57:483–488.
32. Roychowdhury-Saha M, Burke DH. Extraordinary rates of transition metal ion-mediated ribozyme catalysis. *RNA.* 2006;12:1846–1852.
33. Roychowdhury-Saha M, Burke DH. Distinct reaction pathway promoted by non-divalent-metal cations in a tertiary stabilized hammerhead ribozyme. *RNA.* 2007;13: 841–848.
34. Lee T-S, Giambaşu GM, Sosa CP, Martick M, Scott WG, York DM. Threshold occupancy and specific cation binding modes in the hammerhead ribozyme active site are required for active conformation. *J Mol Biol.* 2009;388:195–206.
35. Martick M, Lee T-S, York DM, Scott WG. Solvent structure and hammerhead ribozyme catalysis. *Chem Biol.* 2008;15:332–342.
36. Lee T-S, Silva López C, Giambasu GM, Martick M, Scott WG, York DM. Role of Mg^{2+} in hammerhead ribozyme catalysis from molecular simulation. *J Am Chem Soc.* 2008;130:3053–3064.
37. Lee TS, Silva-López C, Martick M, Scott WG, York DM. Insight into the role of Mg^{2+} in hammerhead ribozyme catalysis from X-ray crystallography and molecular dynamics simulation. *J Chem Theory Comput.* 2007;3:325–327.
38. Anderson M, Schultz EP, Martick M, Scott WG. Active-site monovalent cations revealed in a 1.55-Å-resolution hammerhead ribozyme structure. *J Mol Biol.* 2013;425:3790–3798.
39. Chen H, Giese TJ, Golden BL, York DM. Divalent metal ion activation of a guanine general base in the hammerhead ribozyme: insights from molecular simulations. *Biochemistry.* 2017;56:2985–2994.
40. Vogt M, Lahiri S, Hoogstraten CG, Britt RD, DeRose VJ. Coordination environment of a site-bound metal ion in the hammerhead ribozyme determined by 15N and 2H ESEEM spectroscopy. *J Am Chem Soc.* 2006;128:16764–16770.
41. Kim N-K, Murali A, DeRose VJ. Separate metal requirements for loop interactions and catalysis in the extended hammerhead ribozyme. *J Am Chem Soc.* 2005;127: 14134–14135.
42. Osborne EM, Schaak JE, Derose VJ. Characterization of a native hammerhead ribozyme derived from schistosomes. *RNA.* 2005;11:187–196.
43. Murray JB, Seyhan AA, Walter NG, Burke JM, Scott WG. The hammerhead, hairpin and VS ribozymes are catalytically proficient in monovalent cations alone. *Chem Biol.* 1998;5:587–595.

INDEX

Note: Page numbers followed by "*f*" indicate figures, and "*t*" indicate tables.

A

Acid–base catalysis, 182–191
Aconitase, 112–113
Activated protein C (APC), 46
Adeno-associated viral vector (AAVV), 163–165, 163*t*
Adenoviridae, 166
Adiabatic morphing procedure, 192, 193*f*
Adrenal AT_1Rs, 41–42
Allen Brain Human Atlas, 5–7, 6*f*
α adrenergic receptors (αARs), 29–31
Angiotensin II (AngII) type 1 receptors (AT_1Rs), 28
  adrenal, 41–42
  cardiac, 37–39
  vascular, 39–41
Antibiotic resistance (ABR), 61. *See also* Ciprofloxacin
Antisense region, 85
Apelin receptor, 46
Aquatic hazard assessments, 62–63
Arginine vasopressin (AVP), 43
Arrestin, 28

B

β-adrenergic receptors (βARs), 28, 31–36
βarrestins
  and adrenergic receptors
    α_1ARs, 29–30
    α_2ARs, 31
    β_1ARs, 31–36
    β_2ARs, 36–37
  and angiotensin II receptors
    adrenal AT_1Rs, 41–42
    cardiac AT_1Rs, 37–39
    vascular AT_1Rs, 39–41
  future aspects, 49–50
  and G protein-coupled receptors
    apelin receptor, 46
    endothelin (ET)-1 type A, 42–43
    GPR109A, 43–44
    protease-activated receptor, 45–46
    P2Y family, 44–45
    sphingosine-1-phosphate, 46–47
    vasopressin receptors, 43
  in vivo cardiovascular effects, 32–34*t*
  therapeutic implications, 47–48
Biased signaling, 28–29, 46

C

Carboxypeptidase E (CPE), 4, 5*f*
Cardiac fibroblast-residing V_1R, 43
Cardiovascular GPCRs, 28–29, 42–47
Carvedilol, 29–30
Cell-penetrating peptides (CPP), 167–168
Ciprofloxacin
  aquatic hazard assessments, 62–63
  in aquatic matrices, 63–64, 64*f*
    groundwater examination, 74
    inland freshwaters, 72, 73*f*
    marine and coastal systems, 72–74
    municipal and hospital sewage, 64–72
    wastewater effluent, 66
  discovery, 61
  literature search, 62
Clustered regularly interspaced short palindromic repeats (CRISPR), 80–81
  advantages, 161–162
  electroporation, 159*t*, 168–169
  engineered nucleases, 158, 159*t*
  future aspects, 170–171
  genome silencing *vs.* genome editing, 161
  non-viral vector systems
    advantages, 167
    cationic vectors, 167
    cell-penetrating peptides, 168
    hydrodynamic injections, 168
  opportunities and challenges, 169–170
  programmable nucleases, 158–161
  viral-mediated delivery system
    adeno-associated viral vector, 163–165, 163*t*
    Adenoviridae, 166
    lentiviral vector, 163*t*, 165–166
CsrA protein, 111

D

DEAD-box RhlB helicase, 112
Dead Cas9 (dCas9), 161, 165–166
Degradosome complex, 112–113
Direct genome targeting approach, 80–81
DNA double-strand breaks (DSBs), 80–81, 158

E

EcRNase II, 106–107*f*, 135
Electroporation, 159*t*, 168–169
Endoribonucleases, 102
Endothelin (ET)-1 type A (ET_A) receptors, 42–43
Enolase, 112
Eukaryotic PNPase, 117–119
3′–5′ Exoribonucleases. *See also* Polynucleotide phosphorylase (PNPase); Ribonuclease II (RNase II)
  in microorganisms
    comparative analysis, 103–105, 104*f*
    phylogenetic tree, 103–105, 103*f*
  substrate affinity of *E. coli*, 106–107*f*, 138
Exosome, 123–124
Extended guide sequence (EGS), 85

F

Fast minimal hammerhead ribozyme, 181–182, 183*f*
Fetal gene program, 29–30
Frank–Starling mechanism, 39
Full-length hammerhead ribozyme, 179–181, 180*f*

G

Gene therapy, 80
GPR109A receptor, 43–44
G protein-coupled receptor 83 (GPR83)
  behavioral effects, 12–13
  biological role, 2
  deorphanization, 18–19
  discovery, 3–4
  future aspects, 19
  in hypothalamic function, 12
  immune function, 15–16
  mRNA expression
    in mouse brain, 7–10
    in rat and human brain, 10–11
    regulation, 12
  proSAAS
    discovery, 4
    expression and function, 5–7
    proteolytic processing, 4, 5*f*
  therapeutic effects, 20
G-protein coupled receptors (GPCRs), 2, 42–47
Group I intron
  self-splicing activity, 82–84
  Tetrahymena thermophila, 82, 83*f*
  trans-splicing, 84–85
    for RNA repair, 86–87
    for RNA reprogramming, 87–93
Guide RNA (gRNA), 160*f*, 163–165

H

Hairpin ribozyme, 191
Hammerhead ribozyme, 178
  acid–base catalysis, 179–181, 180*f*
    activator and inhibitor, 183–184
    hairpin ribozyme cleavage reaction, 189–191, 190*f*
    2′-OH of G8 contribution, 187–188, 187*f*
    reverse cleavage mechanism, 189*f*
    RNase A cleavage, 182–183, 185–186, 185*f*
  cleavage mechanism
    base catalysis, 198–200
    ligation, 197–198
  cooperative interactions, 194–195
  fast minimal, 181–182, 183*f*
  ligation reactions, 191–194
  minimal and full-length, 178, 179–180*f*
  structural principles, 196
Hfq, 110, 113, 117
Human PNPase (hPNPase), 110, 117–118
Human telomerase reverse transcriptase (hTERT) RNA reprograming, 89–90
Hydrodynamic injections, 168

L

Lentiviral vector, 163*t*, 165–166

M

Mammillary nuclei, 10
Measured environmental concentration (MEC), 62–63, 67–68*t*, 71*f*
Microscopic Reversibility, 197
Minimal hammerhead sequence, 178, 179*f*
Mitochondrial degradosome complex (mtEXO), 135–136

N

Niacin (nicotinic acid), 43–44
noncoding RNA (ncRNA)
 PNPase activity, 116–117
 RNase II activity, 126
 RNase R activity, 133–135
Non-homologous end joining (NHEJ), 80–81, 109–110

P

Parsimony, 197
PEN, 2, 4–5
PfRNase II, 137
Platelet βarrestin1, 44
Polyethyleneimine (PEI), 167
Polynucleotide phosphorylase (PNPase)
 activity in cells, 108
  DNA repair, 109–110
  regulation by Hfq, 110
  RNA degradation, 108–109
 in eukaryotes, 117–119
 expression, regulators of, 110–112
 homology analysis, 103–105, 104*f*
 multiprotein complexes, 112–113
 pathogenicity functions, 119–121
 RNA metabolism
  coding RNA, 115–116
  noncoding RNA, 116–117
 structure, 113–115
Positron emission tomography-computed tomography (PET-CT), 87–88
Predicted no effect concentrations (PNEC), 61–62, 64–66, 69–70*t*, 71*f*
proSAAS
 discovery, 4
 expression and function, 5–7
 proteolytic processing, 4, 5*f*
Protease-activated receptor (PAR), 45–46

R

Ribonuclease II (RNase II), 103–105
 acetylation, 122–123
 activity, 122
 domain organization and crystal structures, 106–107*f*
 in eukaryotes, 135–136
 expression, 123
 multiprotein complex, 123–124
 pathogenicity functions, 136–138
 RNA metabolism
  coding RNA, 125–126
  noncoding RNA, 126
 structure, 124–125
Ribonucleases, 102
Ribozymes, 81–82. *See also* Hammerhead ribozyme; *trans*-splicing ribozyme
 RNA reprogramming, 87–88
 therapeutic application, 86–87
RNA decay, 102–103
RNA interference (RNAi), 161
RNase A, 182–183
RNase R, 103–105, 122
 in eukaryotes, 135–136
 metabolism
  of coding RNA, 132–133
  of noncoding RNA, 133–135
 multiprotein complexes, 129
 pathogenicity functions, 136–138
 quality control enzyme, 126–127
 RNA degradation, 127–128
 stability, regulation of, 128–129
 structure, 129–132
RNB domain, 103–105, 124–125
RNR1, 117–118, 136
Rrp44/Dis3, 103–105, 122, 131–132

S

Schistosomal hammerhead RNA sequence, 200
Seven transmembrane-spanning receptors (7TMRs), 28
Signal transducer, 41–42, 45
Sphingosine-1-phosphate (S1P), 46–47
SraG, 112
Suppressor cells, 15–16

T

Tetrahymena thermophila, 82, 83*f*
Transcription activator-like effector nucleases (TALENs), 158–161
trans-splicing ribozyme
  development of, 84–85
  molecular imaging, 86
  RNA reprogramming
    cancer cells, 89
    dengue virus, 88–89
    hepatitis C virus, 88–89
    human telomerase reverse transcriptase, 89–90
    mouse CKAP2, 92–93
    mTERT-specific, 92
    tumor-specific, 91–92

V

Vascular AT_1Rs, 39–41
Vascular endothelial cells, 46–47
Vascular smooth muscle (VSM), 29–30, 36–37
Vasopressin receptors, 43

Z

Zinc finger nucleases (ZFNs), 158–161
Zinc finger proteins (ZFPs), 158–161

MIX
Papier | Fördert gute Waldnutzung
FSC® C083411

Kolibri 360 GmbH
Ferdinand-Jühlke-Straße 7
99095 Erfurt, Deutschland
produktsicherheit@kolibri360.de